Managing
Contaminated Sites

Managing Contaminated Sites

Problem Diagnosis and Development of Site Restoration

D. Kofi Asante-Duah

*Environmental Consulting Engineer/Scientist,
Santa Barbara, California, USA
and
Research Assistant Professor, Center for Environmental Engineering and
Science Technologies, University of Massachusetts–Lowell, USA*

JOHN WILEY & SONS

Chichester · New York · Weinheim · Brisbane · Toronto · Singapore

Other Wiley Editorial Offices

John Wiley & Sons, Inc., 605 Third Avenue,
New York, NY 10158-0012, USA

Jacaranda Wiley Ltd, 33 Park Road, Milton,
Queensland 4064, Australia

VCH Verlagsgesellschaft mbH, Pappelallee 3,
D-69469 Weinheim, Germany

John Wiley & Sons (Canada) Ltd, 22 Worcester Road,
Rexdale, Ontario M9W 1L1, Canada

John Wiley & Sons (Asia) Pte Ltd, 2 Clementi Loop #02-01,
Jin Xing Distripark, Singapore 129809

Library of Congress Cataloging-in-Publication Data
Asante-Duah, D. Kofi.
    Managing contaminated sites: problem diagnosis and development of
site restoration/D. Kofi Asante-Duah.
        p.   cm.
    Includes bibliographical references and index.
    ISBN 0-471-96633-9
    1. Hazardous waste site remediation.   I. Title.
TD1030.A83   1996   96-25954
628.5—dc20   CIP

British Library Cataloguing in Publication Data

A catalogue record for this book is available from the British Library

ISBN 0-471-96633-9

Typeset in 10/12pt Times by Dobbie Typesetting Ltd, Tavistock, Devon.
Printed and bound in Great Britain by Bookcraft (Bath) Ltd
This book is printed on acid-free paper responsibly manufactured from sustainable forestation,
for which at least two trees are planted for each one used for paper production.

To all my families at Abaam, Kade, and Nkwantanang
To my mom: Alice Adwoa Twumwaa
To my dad: George Kwabena Duah
To all the Duah brothers and sisters
To all the good friends

Contents

Author biosketch

D. Kofi Asante-Duah, PhD, CE, is an environmental consulting engineer/scientist at Santa Barbara, California, USA, a research assistant professor at the Center for Environmental Engineering and Science Technologies, University of Massachusetts–Lowell, USA, and a member of the Institute for Risk Research, University of Waterloo, Ontario, Canada.

Dr Asante-Duah has served internationally in various capacities (including senior project engineer, project manager, principal investigator, and consultant) for several projects. He has several years of diversified experience with consulting firms and research institutions. His fields of expertise span several topics in the areas of environmental health studies; hazardous waste risk assessment and risk management; design of corrective action programs, including the development of cleanup criteria for site remediation; probabilistic risk assessment and dam safety evaluation; stochastic simulation model applications for solving water management problems; decision analysis approaches to environmental management; statistical evaluation of environmental data; environmental impact assessment; and environmental policy analyses. He has worked on projects relating to leachate migration from landfills and other contaminated sites into aquifers; modeling surface water contamination from hazardous waste sites; review and evaluation of remedial action plans for inactive waste sites, culminating in recommendations for site remediation; and use of parametric statistical analyses to determine impacts of industrial discharges into streams. Dr Asante-Duah's projects have also included studies of the effects of household hazardous wastes on human health and the environment, based on evaluation of alternative disposal practices; comparative evaluation of alternative remedial options for industrial facilities; policy analyses of the tradeoffs involved in the transboundary movements of hazardous wastes; risk management and prevention programs for facilities handling acutely hazardous materials; and human health and environmental risk assessments for industrial facilities and contaminated sites.

Dr Asante-Duah has previously been on visiting appointments to the University of Pittsburgh, USA, Civil Engineering Department (as visiting scholar/scientist in 1985/86) and the University of Waterloo, Canada, Civil Engineering Department (as visiting research assistant professor in 1990/91). He has previously been a research assistant at the Utah Water Research Laboratory, Utah State University, USA (1986–88). Dr Asante-Duah is a past UNESCO fellow (1984) and also a World Bank's McNamara fellow (1990–91). He has affiliations with several international professional associations.

Dr Asante-Duah is the author or coauthor of several technical publications and presentations relating to water resources evaluation, risk assessment, and hazardous waste management.

Preface

Site contamination by a variety of toxic chemicals has become a major environmental issue in several industrialized and industralizing countries. Manufacturing and related activities apparently are responsible for many environmental contamination and contaminated site problems. Irrespective of the contributing source(s), however, once a contaminant is released into the environment, it may be transported and/or transformed in a multiplicity of complex ways. These types of situations have, in several ways, contributed to the widespread occurrence of contaminated sites and related environmental contamination problems globally.

Despite the variability in the nature of contaminated site problems encountered at different locations, there seems to be growing consensus on the general technical principles and procedures that should be employed in the management of such sites. This book attempts to synthesize and summarize the relevant *generic principles* for application to the variety of contaminated site situations often encountered in practice. The book brings together the common themes and concepts that are applicable to making credible decisions about the reclamation of contaminated lands for re-development. It specifically elaborates on the scientific and technical requirements for contaminated site assessment and site restoration efforts.

In order to be able to develop an effectual corrective action program, contaminated sites have to be extensively studied to determine the real extent of contamination, the quantities of contaminants that human and ecological receptors could potentially be exposed to, the health and ecological risks associated with the site, and the types of corrective or remedial actions necessary to abate risks from such sites. The book offers an elaboration of the requisite procedures utilized in the planning, development, and evaluation of corrective action assessment and response programs for such impacted sites. The book focuses on the principles involved in the effective management of contaminated sites, accomplished via the use of appropriate diagnostic tools to determine the degree of contamination and/or the risk, and the subsequent development of an effectual site restoration program.

This book represents a collection and synthesis of technically sound principles and concepts that should find unbiased global application to contaminated site problems. It is expected to serve as a useful *educational and training resource for both students and professional consultants* dealing with contaminated land assessment and reclamation issues. In fact, the fundamental technical principles and concepts involved in the management of contaminated site problems will generally not differ in any significant way from one geographical region to another. Consequently, this book will serve as useful reference material for the global community dealing with contaminated site assessment and restoration problems. The systematic protocols presented will aid several environmental professionals to formulate and manage contaminated sites and associated problems more efficiently. It is the hope of the author that the three-part presentation offered by this title will give adequate guidance and direction for the successful completion of corrective action assessment and response programs that are to be designed for any type of contaminated site problem, and at any geographical location.

I am indebted to a number of people for both the direct and indirect support afforded me during the preparation of this book. Sincere thanks are due the Duah family (of Abaam, Kade, and Nkwantanang), and several friends and colleagues who provided much-needed moral and enthusiastic support throughout the preparation of the manuscript for this book. I thank Dr Hilary I. Inyang (Director, Center for Environmental Engineering and Science Technologies, University of Massachusetts–Lowell, USA), for being supportive in so many ways, and the Publishing, Editorial and Production staff at John Wiley & Sons (Chichester, England) who helped bring this book project to a successful completion. I also wish to thank the numerous authors whose work are cited in this volume, for having provided some pioneering work to build on. Review comments and suggestions on an earlier draft of the manuscript for this book was provided by Dr L. Douglas James, National Science Foundation, Washington, DC, USA. This book also benefited from review comments of several anonymous individuals, as well as from discussions with a number of professional colleagues. Any shortcomings that remain are, however, the sole responsibility of the author.

D. Kofi Asante-Duah
November 1995

PART I

GENERAL OVERVIEW

Chapter One

Introduction

Site contamination by a variety of toxic chemicals has become a major environmental issue in several industrialized and industrializing countries. Industrial activities and related waste management facilities apparently are responsible for many environmental contamination and contaminated site problems. The contributing waste management activities may relate to industrial wastewater impoundments, land disposal sites for solid wastes, land spreading of sludges, accidental chemical spills, leaks from chemical storage tanks and piping systems, septic tanks and cesspools, disposal of mine wastes, or indeed a variety of waste treatment, storage and disposal (TSD) facilities. Once a chemical constituent is released into the environment, it may be transported and/or transformed in a variety of complex ways. These types of situations have, in several ways, contributed to the widespread occurrence of contaminated sites and related environmental contamination problems globally.

This book will focus on the principles involved in the effective management of contaminated sites via the use of appropriate diagnostic tools to determine the degree of contamination, hazard, and/or risk, and the subsequent development of effectual site restoration programs. The major features of the elaboration in the book are annotated in Box 1.1. Overall, the book provides a wealth of information which can be applied to site appraisal and assessment, and to the implementation of contaminated site restoration programs. In any case, to be able to develop an effectual corrective action program, contaminated sites have to be extensively studied to determine the areal extent of contamination, the quantities of contaminants that human and ecological receptors could potentially be exposed to, the human health and ecological risks associated with the site, and the types of corrective or remedial actions necessary to abate risks from such sites.

1.1 THE BIRTH OF CONTAMINATED SITES

Site contamination occurs when chemicals are detected where such constituents are not expected and/or not desired. Contaminated sites may arise in a number of ways, many of which are the result of manufacturing and other industrial activities or operations. In fact, many of the environmental contamination problems encountered in a number of places are the result of waste generation associated with various forms of industrial activities. Wastes are generated from several operations associated with industrial (e.g. manufacturing and mining), agricultural, military, commercial (e.g. automotive repair shops, utility companies, fueling stations, dry-cleaning facilities, transportation centers

---

Box 1.1   Major features of the book

- It brings together the common themes and concepts that are applicable to making credible decisions about the reclamation of contaminated lands for re-development.
- It synthesizes and summarizes the relevant generic principles for application to the wide variety of contaminated site situations often encountered in practice.
- It presents concepts generally applicable to the broad spectrum of environmental contamination problems.
- It elaborates on the scientific and technical requirements for contaminated site assessment and site restoration efforts.
- It offers an elaboration of the procedures utilized for planning, developing, and evaluating corrective action programs for potentially contaminated sites and facilities.
- It discusses strategies for the effective management of contaminated site problems, using effectual corrective action programs, in order to abate the potential risks associated with such sites.
- It addresses issues relevant to the investigation and the management of potentially contaminated site problems, spanning problem diagnosis and site characterization to the development and implementation of site restoration programs.
- It presents risk-based technical methods to facilitate the development of cost-effective diagnostic and site restoration decisions for contaminated site problems.

---

and food processing industries) and domestic activities. In particular, the chemicals and allied products manufacturers are generally seen as the major sources of industrial hazardous waste generation. Table 1.1 contains a summary of typical industries generating large volumes of hazardous wastes that could be contributing to the births of contaminated sites. These industries generate several waste types, such as organic waste sludges and still bottoms (containing chlorinated solvents, metals, oils, etc.); oil and grease (contaminated with polychlorinated biphenyls (PCBs), polyaromatic hydrocarbons (PAHs), metals, etc.); heavy metal solutions (of arsenic, cadmium, chromium, lead, mercury, etc.); pesticide and herbicide wastes; anion complexes (containing cadmium, copper, nickel, zinc, etc.); paint and organic residuals; and several other chemicals and byproducts that need special handling/management.

Contaminated site problems typically are the result of soil contamination due to placement of wastes on or in the ground; as a result of accidental spills, lagoon failures or contaminated runoff; and/or from leachate generation and migration. Contaminants released to the environment are affected by a complex set of processes that include various forms of transport and cross-media transfers, transformation, and biological uptake. For instance, atmospheric contamination may result from emissions of contaminated fugitive dusts and volatilization of chemicals present in soils; surface water contamination may result from contaminated runoff and overland flow of chemicals (from leaks, spills, etc.) and chemicals adsorbed on to mobile sediments; groundwater contamination may result from the leaching of toxic chemicals from contaminated soils or the downward migration of chemicals from lagoons and ponds. Indeed, several different physical and chemical processes can affect contaminant migration from a contaminated site, as well as the cross-media transfer of contaminants at any given site. Consequently, contaminated soils can potentially impact several other environmental matrices.

Table 1.1 Major industries and manufacturers potentially contributing to contaminated site problems

- Aerospace
- Ammunitions
- Automobile
- Batteries
- Beverages
- Chemical production
- Computer manufacture
- Electronics and electrical
- Electroplating and metal finishing
- Explosives manufacture
- Food and dairy products
- Herbicides, insecticides, and pesticides
- Ink formulation
- Inorganic pigments
- Iron and steel
- Leather tanning and finishing
- Metal smelting and refining
- Mineral exploration and mining
- Paint products
- Perfumes and cosmetics
- Petroleum products
- Pharmaceutical products
- Photographic materials
- Printing and publishing
- Pulp and paper mills
- Rubber products, plastic materials, and synthetics
- Shipbuilding
- Soap and detergent manufacture
- Textile products
- Wood processing and preservation

1.2 A CONTAMINATED SITE CLASSIFICATION SYSTEM

Contaminated sites may pose different levels of risk, depending on the nature and extent of contamination present at the site. The degree of hazard posed by the contaminants involved will generally be dependent on several factors such as: physical form and composition of contaminants, quantities of contaminants, reactivity, biological and ecological toxicity effects, mobility in various environmental media, persistence or attenuation in environment, and local site conditions (e.g. temperature; soil type; groundwater flow conditions; humidity; and light).

In fact, it is important to recognize the fact that there may be varying degrees of hazards associated with different contaminated site problems, and that there are good technical and economic advantages for ranking potentially contaminated sites according to the level of hazard they present. A typical site categorization scheme for potentially contaminated sites will comprise of putting the 'candidate' sites into groups or clusters, based on the potential risks associated with such sites (e.g. high-, intermediate- and low-risk sites, conceptually represented by Figure 1.1). Such a

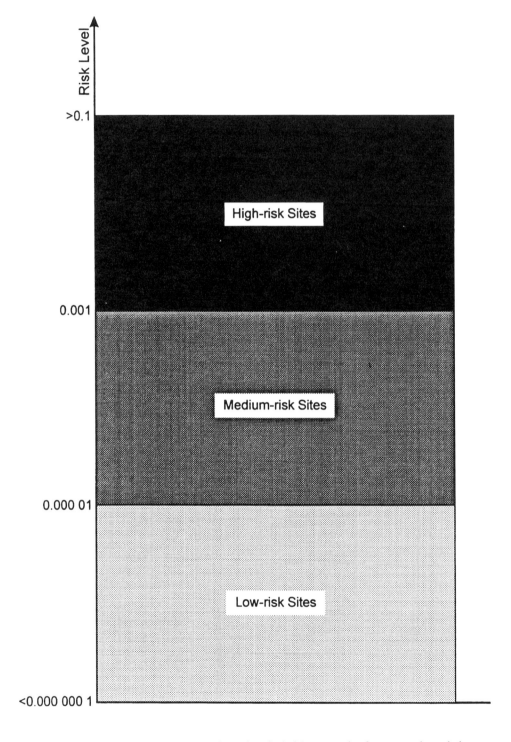

Figure 1.1   A conceptual representation of typical risk categories for contaminated sites

classification will facilitate the development and implementation of efficient site management or restoration programs.

In general, the high-risk sites will prompt the most concern, requiring immediate and urgent corrective measures that may include time-critical removal actions. A site is designated as high-risk when site contamination represents a real or imminent threat to human health and/or to the environment. In this case, an immediate action will generally be required to reduce the threat. Thus, in order to ensure the development of adequate site management and restoration strategies, potentially contaminated sites should preferably be categorized in an appropriate manner.

1.3 HEALTH AND ENVIRONMENTAL EFFECTS FROM CONTAMINATED SITES

The mere existence of contaminated sites can lead to contaminant releases and possible receptor exposures, resulting in both short- and long-term effects on a variety of populations potentially at risk. Lessons from the past, as recorded in several US locations, in Europe, in Japan, and in other places in Asia, clearly demonstrate the dangers that may result from the presence of contaminated sites within or near residential communities (Asante-Duah, 1993; Brooks et al., 1995; Grisham, 1986). Exposure to chemical constituents present at contaminated sites can indeed produce adverse effects in both human and ecological receptors. For example, human exposures to certain environmental contaminants may result in such diseases as allergic reaction, anemia, anxiety, asthma, blindness, bronchitis, various cancers, contact dermatitis, convulsions, embryotoxicity, emphysema, heart disease, hepatitis, obstructive lung disease, memory impairment, nephritis, neuropathy, and pneumoconiosis; ecological exposures to certain chemicals may result in such toxic manifestations as bioaccumulation and/or biomagnification in aquatic organisms. In general, any chemical present at a contaminated site can cause severe health impairment or even death if taken by organisms (including humans) in sufficiently large amounts. On the other hand, there are those chemicals of primary concern which can cause adverse impacts, even from limited exposures.

The potential for adverse health effects on populations contacting hazardous chemicals present at a contaminated site can involve any organ system. The target and/ or affected organ(s) will depend on the specific chemicals contacted; the extent of exposure (i.e. dose or intake); the characteristics of the exposed individual (e.g. age, gender, body weight, psychological status, genetic make-up, immunological status, susceptibility to toxins, hypersensitivities); the metabolism of the chemicals involved; weather conditions (e.g. temperature, humidity, barometric pressure, season); and the presence or absence of confounding variables such as other diseases (Brooks et al., 1995; Grisham, 1986).

1.3.1 Human Health and Ecologic Effects of Environmental Contaminants

Several health effects may arise when human and ecological receptors are exposed to some agent or stressor present in the environment. The following represent the major categories of human health and ecologic effects that could result from exposure to environmental contaminants (Brooks et al., 1995; Grisham, 1986):

- Human health
  - Carcinogenicity (i.e. capable of causing cancer in humans and/or laboratory animals).
  - Heritable genetic and chromosomal mutation (i.e. capable of causing mutations in genes and chromosomes that will be passed on to the next generation).
  - Developmental toxicity and teratogenesis (i.e. capable of causing birth defects or miscarriages, or damage to developing fetus).
  - Reproductive toxicity (i.e. capable of damaging the ability to reproduce).
  - Acute toxicity (i.e. capable of causing death from even short-term exposures).
  - Chronic toxicity (i.e. capable of causing long-term damage other than cancer).
  - Neurotoxicity (i.e. capable of causing harm to the nervous system).
- Ecologic
  - Environmental toxicity (i.e. capable of causing harm to wildlife and vegetation).
  - Persistence (i.e. does not break down easily, thus persisting and accumulating in portions of the environment).
  - Bioaccumulation (i.e. can enter the bodies of plants and animals but is not easily expelled, thus accumulating over time through repeated exposures).

A number of chemicals encountered at contaminated sites are known or suspected to cause cancer; several others may not have carcinogenic properties, but are nonetheless of significant concern due to their systemic toxicity effects. In fact, several different symptoms, health effects and other biological responses may be produced from exposure to various specific toxic chemicals commonly encountered at contaminated sites. The presence of contaminated sites can therefore create potentially hazardous situations and pose significant risks of concern to society.

Invariably, exposures to chemicals escaping into the environment can result in a reduction of life-expectancy and possibly a period of reduced quality of life. The existence of unregulated contaminated sites in society can therefore be perceived as a potential source of several health, environmental, and possibly socio-economic problems.

1.4   RESTORING THE INTEGRITY OF CONTAMINATED SITES

To avert possible detrimental effects to human health and the environment, an attempt is often made to restore the integrity of contaminated sites. In this regard, corrective actions for contaminated sites are generally developed and implemented with the principal objective to protect public health and the ecosystem. Typically, answers will have to be generated for several pertinent questions when one is confronted with a potentially contaminated site problem. In particular, answers to the following questions will help define the corrective action needs for the case site:

- What is the nature of contamination?
- What are the sources of, and the 'sinks' for, the site contamination?
- What is the current extent of contamination?

- What population groups are potentially at risk?
- What are the likely and significant exposure pathways and scenarios representative of the site?
- What is the likelihood of health and environmental effects resulting from the contamination?
- What interim measures, if any, are required as part of a risk management and prevention program?
- What corrective action(s) may be appropriate to remedy the prevailing situation?
- What level of residual contamination will be acceptable for site restoration?

In a number of situations, it becomes necessary to implement interim corrective measures prior to the development and full implementation of a comprehensive site restoration program. Such preliminary corrective actions may consist of a variety of site control activities, such as the installation of security fences (to restrict access to the site); the construction of physical barriers (to restrict site access and also to minimize potential runoff and run on from the site); the application of dust suppressants (to reduce airborne migration of contaminated soils to off-site locations); the removal of 'hot spots' and drums (where acute toxic exposure could possibly occur), etc. Ultimately, a thorough site investigation that establishes the nature and extent of contamination may become necessary, in order to arrive at an appropriate and realistic site restoration plan.

1.4.1 Regulatory Considerations Affecting Contaminated Sites

Increasing public concern about the several problems and potentially dangerous situations associated with environmental contamination problems, together with the legal provisions of various legislative instruments and regulatory programs, have all compelled both industry and governmental authorities to carefully formulate responsible environmental management programs. These programs include techniques and strategies needed to provide good waste management methods and technologies, and the development of cost-effective corrective action programs that will ensure public safety, as well as protect human health, the environment, and public and private properties.

A general discussion and analysis of hazardous waste management practices in several industrialized countries is presented by Forester and Skinner (1987) in a report of the International Solid Wastes and Public Cleansing Association (ISWA) working group on hazardous wastes. Generally, most industrialized countries incorporate a system for 'cradle-to-grave' control in their waste management programs. The 'cradle-to-grave' type of system monitors and regulates the movement of hazardous materials from manufacture through usage to the ultimate disposal of any associated hazardous wastes. By using some kind of manifest system, this helps minimize abuses and violations of established national or regional control systems associated with hazardous waste movements and management. The manifest system serves as an identification form that accompanies each shipment of wastes; the manifest is signed at each stage of transfer of responsibility (i.e. from the waste generator, to the transporter, to the storage, treatment and/or disposal operator), with each responsible person in the chain of custody keeping a record that is open to scrutiny and inspection by regulatory officials or other appropriate authorities, and also other interested parties. The use of such effective control systems should help minimize the creation of extensive environmental contamination problems.

To help abate potential problems of public health and the environment, several items of legislation have been formulated and implemented in most industrialized countries to deal with the regulation of toxic substances present in our modern societies.

Invariably, there tend to be improved environmental management practices in countries or regions where regulatory programs are well established, in comparison to those without appropriate regulatory and enforcement programs. The process used in the design of effectual contaminated site management programs must therefore incorporate several elements of all relevant environmental regulations. In any case, it is noteworthy that, variations in national legislations and controls do affect options available for the management of contaminated site problems in different regions of the world. Such variances also affect the cleanup standards and costs necessary to achieve the appropriate program goals. The need to consult with local environmental regulations when faced with a contaminated site management problem cannot, therefore, be over-emphasized. In fact, the establishment of national and/or regional regulatory control programs within the appropriate legislative–regulatory and enforcement frameworks is a major step in developing effective waste and contaminated site management programs.

1.5   REFERENCES

Asante-Duah, D.K. (1993). *Hazardous Waste Risk Assessment*. CRC Press/Lewis Publishers, Inc., Boca Raton, Florida.

Brooks, S.M. et al. (1995). *Environmental Medicine*. Mosby, Mosby-Year Book, Inc., St Louis, Missouri.

Forester, W.S. and J.H. Skinner (eds) (1987). *International Perspectives on Hazardous Waste Management*. A Report from the International Solid Wastes and Public Cleansing Association (ISWA) Working Group on Hazardous Wastes. Academic Press, London, UK.

Grisham, J.W. (ed.) (1986). *Health Aspects of the Disposal of Waste Chemicals*. Pergamon Press, Oxford, UK.

PART II
PROBLEM DIAGNOSIS

Chapter Two

Investigating potentially contaminated sites

In order to be able to characterize the nature and extent of suspected contamination, an environmental site investigation is generally required in the study of potentially contaminated site problems. Because of the inherent variability in the materials and the diversity of processes used in industrial activities, it is not unexpected to find a wide variety of environmental contaminants at any particular contaminated site. As a consequence, there is a corresponding variability in the range and type of hazards and risks that may be anticipated from different contaminated site problems. Contaminated sites should therefore be investigated very carefully and thoroughly so that risks to potentially exposed populations can be determined with a high degree of accuracy. Typically, different levels of effort in the investigation will generally be required for different contaminated site problems. Also regulations in different countries or locales will affect the site investigation protocol.

In formulating site investigation programs, it should be recognized that the most important primary sources of contaminant release to the various environmental media are usually associated with constituents in soils. The contaminated soils can subsequently impact other environmental matrices by a variety of processes, as discussed in Chapter 3. The impacted media, having once served as 'sinks', may eventually become secondary sources of contaminant releases into other environmental compartments. In general, all relevant sources and 'sinks' or impacted media should be thoroughly evaluated as part of the site investigation efforts.

2.1 THE SITE INVESTIGATION PROCESS

Site investigations consist of the planned and managed sequence of activities carried out to determine the nature and distribution of contaminants at potentially contaminated sites. The activities involved usually are comprised of the following (BSI 1988):

- Identification of the principal hazards.
- Design of sampling and analysis programs.
- Collection and analysis of environmental samples.
- Recording or reporting of laboratory results for further evaluation.

In order to get the most out of a site investigation, it must be conducted in a systematic manner. Systematic methods help focus the purposes, the required level of detail, and the several topics of interest – such as physical characteristics of the site, likely contaminants, extent and severity of possible contamination, effects of contaminants on populations potentially at risk, probability of harm to human health and the environment, and possible hazards during construction activities (Cairney, 1993). The systematic process required for the site investigation essentially involves the early design of a representative conceptual model of the site, as detailed in Chapter 4. This model is used to assess the physical conditions at the site as well as to identify the mechanisms and processes that could produce significant risks at the site and its vicinity.

2.1.1   A Tiered Approach to Site Investigations

A tiered approach may be adopted in the investigation of potentially contaminated sites, to provide a cost-effective way of determining the true extent of contamination at such a site. The use of a phased approach will generally result in an optimal data gathering and evaluation process that meets requisite data quantity and quality objectives. Consequently, programs designed to investigate potentially contaminated site problems typically consist of a number of phases. These phases, reflecting the different degrees of detail, may be classified into the following successive tiers:

- *Tier 1* – Preliminary Site Appraisal, consisting of a records search together with a reconnaissance site appraisal and reporting.
- *Tier 2* – Primary Site Investigation, comprising of a site investigation that involves contamination and environmental damage assessment.
- *Tier 3* – Expanded Site Investigation, incorporating a comprehensive site assessment that also includes a preliminary feasibility study of site restoration measures.

In general, the objective of the initial phase (which is comprised of basic background information gathering) should be to determine the history of the site with respect to contamination sources and any relevant characteristics of the site that are readily obtainable from historical records, reports, and interviews. The initial phase of investigation commonly may also employ field screening methods to delineate the approximate area and general magnitude of the problem.

The intermediate phase (involving a site characterization) has the primary objectives of defining the vertical and lateral extents of contamination, and also understanding how the contaminants are affected by hydrogeological conditions at the project site. The investigation of hydrogeological conditions at the project site typically involves an evaluation of groundwater quality, groundwater flow directions, thickness and relative locations of aquifers and confining units, and potential discharge points for groundwater such as local abstraction wells or surface water bodies (Pratt, 1993). Such information is essential to the determination of potential migration pathways and exposure routes for contaminants in groundwater, especially in areas where groundwater is a significant source of potable water supply.

The final phase of the site investigation provides the data necessary to design appropriate and applicable site restoration measures.

Typically, successive tiers call for increasingly sophisticated levels of data collection and analysis. The requirements of the different levels of effort associated with the site assessment process are elaborated below.

2.1.1.1   Tier 1 investigations – the preliminary site appraisal

A preliminary site appraisal (PSA) is usually conducted to determine if a site is potentially contaminated as a result of past or current site activities, unauthorized dumping or disposal, or the migration of contaminants from adjacent or nearby properties.

The purpose   The purpose of the PSA is to provide a qualitative indication of potential contamination at a site. The PSA typically is designed to document known or potential areas of concern due to the presence of contaminants, to establish the characteristics of the contaminants, to identify potential migration pathways, and to determine the potential for migration of the contaminant constituents.

The tasks   The PSA will, at a minimum, involve record searching and a superficial physical survey that consist of the following activities:

- Review of historical records such as site history, past operation and disposal practices, nature of chemicals used and waste derivatives, etc.
- Review of readily available files and databases maintained by various regulatory and other governmental agencies, in order to obtain information on site hydrogeology, characteristics of adjacent properties, known environmental problems in the general area, etc.
- Field reconnaissance of the subject site and adjacent properties, in order to ascertain local soil and groundwater conditions, the proximity of the site to drinking water supplies and surface water discharges, existence of obvious contaminant sources or areas, etc.

Data collection for the PSA is generally accomplished by sequentially performing a records search/review, a site reconnaissance, and where practicable and/or appropriate, a soil vapor contaminant assessment. Typical specific tasks performed to meet the overall objective of a PSA will consist of a geologic and hydrogeologic literature search, aerial photo reviews, review of archival and regulatory or governmental agency records (to identify the historical uses of site), and review of anecdotal reports (on site history and practices that are made available by former employees, local residents, and local historians); visual site inspection/reconnaissance survey (to define the present condition of the site); personal interviews (to supplement historical use data); and a written report of findings (to document results and recommendations). After this initial site appraisal, the information so obtained is summarized, and a preliminary sampling and analysis plan is developed, as appropriate.

The results   A Tier 1 investigation will generally conclude that either: (1) *no current or historic evidence of contamination considered likely to have affected the site is present, and therefore no further investigation is required,* or (2) *sources of contamination that may have affected the site have been identified, and therefore further investigation is required at the Tier 2 level.* Thus, depending on the results of the PSA, a 'further-response-action' or a 'no-further-response-action' may be recommended for the project site.

   Whereas the PSA provides some important site information, a more detailed characterization of the extent of soil and groundwater contamination may be necessary to properly assess long-term risks posed by a contaminated site. Consequently, where warranted, the next level of detail in the site investigation or assessment process is carried out to ascertain any initial indication of possible contamination.

2.1.1.2   Tier 2 investigations – the primary site investigation

The primary site investigation (PSI) should be designed to verify findings from the PSA, and further to determine the presence (and the extent) or absence of contamination at the site.

The purpose   The purpose of the PSI is to identify the known or suspected source(s) of contamination, and also to define the nature and extent of the contamination. Typically, the investigation identifies specific contaminants, their concentrations, the areal extent of contamination, the fate and transport properties of the contaminants, and the potential migration pathways of concern.

The tasks   A Tier 2 investigation will normally be used to confirm whether or not a release has occurred. This is accomplished by implementing a limited program to collect and analyze appropriate site samples for some target contaminants. Typical tasks performed during this phase of the site investigation consist of those identified under the PSA, plus the following additional activities:

• Sampling of potentially impacted media at the surface.
• Subsurface borings, well installations, and groundwater sample collection.
• Sample analysis to identify and quantify contaminants.

Subsequently, an evaluation of the sampling results and the preparation of a detailed report of findings, with conclusions, will help document the results of the Tier 2 investigation.

The results   A Tier 2 investigation will generally conclude that either: (1) *no evidence of contamination was discovered, and therefore no further investigation is required*, or (2) *contamination that may require remediation has been found, and therefore a Tier 3 level of investigation is recommended*. Consequently, the results of this initial comprehensive site assessment are used to determine the need for a 'further-response-action' or a 'no-further-response-action'. In fact, if the site is determined to pose significant public health or environmental risks, extensive studies will typically be required to quantify the magnitude of contaminants present, to delineate the limits of contamination, to characterize in detail the specific chemical constituents present at the site, and to assess the fate and transport properties of the specific substances at the site.

2.1.1.3   Tier 3 investigations – the expanded site investigation

The expanded site investigation (ESI) strives to improve the initial site characterization and also to identify, on a preliminary basis, the most cost-effective methods of remediation that will protect public health, the environment, and public and private property under applicable and appropriate cleanup goals or regulations.

The purpose   An ESI is conducted with the main objective to fully characterize the contamination confirmed from the PSI. The overall purpose of the ESI is to define the regional hydrogeology and direction of contaminant plume migration, and also to identify those remedial options potentially applicable to an anticipated site restoration program.

The tasks The ESI process typically involves specifying the type of contamination present, assessing the three-dimensional occurrence of the contamination, evaluating the contaminant fate and transport, identifying possible human and ecological receptors potentially at risk, estimating the risks posed to the populations at risk, establishing a database to facilitate documentation of changes in the occurrences of the contamination, and conducting a preliminary screening of site restoration measures.

At a minimum, the ESI will include the collection and analysis of as many soil and water samples as necessary, in order to determine the full extent of site contamination. If contamination is found to be confined to the unsaturated (vadose) zone, then no groundwater investigation may be necessary; otherwise, groundwater investigations are also carried out. As appropriate, other environmental media (e.g. surface water, sediments, and biota) may also have to be sampled. The level of detail for the data collection activities will be site-specific, and is dependent on the degree of soil and groundwater contamination found at the site. Typical tasks performed during the Tier 3 investigation comprise of those identified under the PSI, plus the following:

- Development of remediation goals and cleanup criteria.
- Identification of alternative methods/technologies for a site restoration.
- Screening of remediation alternatives, to aid in the future selection of those techniques that are most feasible.

In general, comprehensive site assessments usually will involve sampling and testing to identify the types of contaminants, analyzing pre-selected or 'priority' pollutants, and determining the horizontal and vertical extents of the contamination. This typically includes subsurface investigations, soil and water sampling, laboratory analyses, storage tank testing, and other relevant engineering investigations to quantify potential risks previously identified in the PSA and/or PSI.

The results A Tier 3 investigation will generally conclude that either: (1) *no evidence of extensive contamination was discovered that requires remediation, and therefore no further investigation is required,* or (2) *contamination that may require remediation has been found, and therefore a site restoration program should be developed and implemented.* Typically, the Tier 3 assessment will generate a report made up of a site characterization, a risk assessment, an evaluation of mitigation and remediation options, and an indication of the preferred remedial action plan.

2.2 ELEMENTS OF A SITE INVESTIGATION PROGRAM

In the course of investigating potentially contaminated sites, data quality objectives (DQOs) are used as qualitative and quantitative statements that specify the nature, extent and quality of data required to support site restoration decisions. (Restoration decisions have been limited by too little data in the past.) Overall, the DQO process results in a well thought out sampling and analysis plan. Consequently, DQOs should preferably be established prior to data collection activities, to ensure that the data collected are sufficient and of adequate quality for their intended uses. The DQOs

Table 2.1   Tasks and important elements of a site investigation program

Task	Elements
Problem definition	• Define project objectives (including the level of detail and topics of interest) • Determine data quality objectives (DQOs)
Preliminary evaluation	• Collect and analyze existing information (i.e. review available background information, previous reports, etc.) • Conduct visual inspection of the site (i.e. field reconnaissance surveys) • Construct preliminary conceptual model of site
Sampling design	• Identify information required to refine conceptual model of the site • Identify constraints and limitations (e.g. access, regulatory controls, utilities, and financial limitations) • Define sampling, analysis, and interpretation strategies • Determine exploratory techniques and testing program
Implementation of sampling and analysis plans	• Conduct exploratory work on site (e.g. exploratory borings, test pits, and geophysical surveys) • Perform *in-situ* testing, as appropriate • Carry out sampling activities • Compile record of investigation logs, photographs, and sample details • Perform laboratory analyses
Data evaluation	• Compile a database of relevant site information • Carry out logical analysis of site data • Refine conceptual model for site
Interpretation of results	• Enumerate implications of site investigation results • Prepare a report on findings

should subsequently be integrated into the development of sampling and analysis plans, and should be revised as needed based on the results of each data collection activity.

Oftentimes, site investigation activities are designed and implemented in accordance with several regulatory and legal requirements of the geographical region or area in which a potentially contaminated site or property is located. In a typical investigation, the influence of the responsible regulatory agencies may affect the several operational elements (elaborated in Chapter 5) necessary to complete the site investigation program. Irrespective of whichever regulatory authority is involved, however, the basic site investigation strategy generally adopted for contaminated site problems typically will comprise of the specific tasks and general elements summarized in Table 2.1 (Cairney, 1993; USEPA, 1988b). Ultimately, the data derived from the site investigation may be used to perform a risk assessment (outlined in Chapter 6) that becomes a very important element in the site restoration decision.

2.3   DATA COLLECTION AND ANALYSIS

The design and implementation of a substantive data collection and evaluation program is vital to the effective management of contaminated sites. Data are generally collected at several stages of the site investigation, with initial data collection efforts usually limited to developing a general understanding of the site. The general types of site data and information required in the investigation of contaminated sites relate to the following (USEPA, 1989a):

- Contaminant identities.
- Contaminant concentrations in the key sources and media of interest.
- Characteristics of sources and contaminant release potential.
- Characteristics of the physical and environmental setting that can affect the fate, transport, and persistence of the contaminants.

In general, a detailed records search of background information on the critical contaminants of potential concern should be compiled as part of the contaminated site investigation program. The investigation must then provide information on all contaminants known, suspected or believed to be present at the site. In fact, the investigation should cover all compounds for which the history of site activities, current visible contamination, or public concerns suggest that such substances could be present. In addition to establishing the concentration of contaminants at a site, the site investigation should be designed to provide an indication of the naturally-occurring or anthropogenic background level of the target contaminants in the local environment.

Ultimately, several chemical-specific factors (such as toxicity or potency, concentration, mobility, persistence, bioaccumulative or bioconcentration potential, synergistic or antagonistic effects, potentiation or neutralizing effects, frequency of detection, and naturally-occurring or anthropogenic background thresholds) are used to screen further and select the specific target contaminants that will become the focus of a detailed site evaluation process. Typically, the selected target chemicals are those site contaminants that are generally the most mobile and persistent; consequently, they reflect the likelihood of contamination at the site.

2.3.1 The General Types of Field Investigations

Field studies are necessarily iterative processes, beginning with activities to determine both surface and subsurface conditions at a project site. Information collected from a review of the historical records of a potentially contaminated site, and also the current operating conditions, can be used to develop a general approach to the data collection and analysis process. A variety of selected field investigation methodologies that may be employed during the investigation of contaminated site problems are enumerated below.

Geophysical surveys The use of geophysical methods in general, and surface geophysical techniques in particular, can significantly reduce the amount of time and the cost involved in site characterization activities. It is therefore considered good practice to use surface geophysical surveys wherever feasible in the investigation of potentially contaminated site problems. For instance, geophysical survey techniques may be used: to determine the lateral extent of past landfilling activities; to define areas having extensive disturbed soil formations; to assess the presence of unknown buried metallic objects or voids; and to demarcate and clear utilities so that drilling activities may proceed. These techniques may also assist in the placement of groundwater monitoring wells, as well as in estimating the existence of preferential groundwater flow directions.

Overall, surface geophysical surveys are useful for providing information on subsurface geologic features in areas where limited stratigraphic data exist. When performed properly and utilized early in the site characterization process, surface geophysics can provide valuable information to use in planning monitoring well and piezometer placement. In addition, surface geophysics can be used to correlate the stratigraphy and hydrostratigraphy between wells, locate buried structures, and in some instances, can directly detect underground contaminant plumes.

The more commonly used geophysical survey methods employed in the investigation of contaminated sites include ground-penetrating radar (GPR), pipe and cable locator, electromagnetic (EM) surveys (using terrain conductivity meters), magnetometer, and geophysical gamma logging in boreholes (CDHS, 1990; Osiensky, 1995; Telford et al., 1976; USEPA, 1985). Indeed, borehole geophysical techniques can be of significant interest in their ability to provide efficient and cost-effective means of collecting lithologic and hydrologic information from wells and borings. These methods provide continuous measurements of physical properties along the entire length of a borehole, supplementing the discrete information gathered by coring.

Trial pitting   Trial pitting using a back-hoe excavator is probably the most common technique for investigating the condition of shallow subsurface soil (Pratt, 1993). Typically, visual evidence of contamination (such as apparent staining or the presence of landfilled materials) as well as odors can be used to guide the selection of sampling location, where samples are to be collected for laboratory analyses.

Soil gas surveys   In situations where organic chemicals represent the major source of contamination, a preliminary gas survey using subsurface probes and portable equipment can give an early indication of likely problem areas. Soil gas surveys are generally carried out as a precursor to exploratory excavations, in order to identify areas that warrant closer scrutiny. They can also be used to assist in the delineation of previously identified plumes of contamination. This can be a very important step to complete prior to the start of a full-scale site investigation that employs intrusive exploratory methods. Soil gas surveys can also be used to evaluate and identify large areas with organic contamination in a relatively short period of time (Pratt, 1993).

For a contaminant to be identified in a soil gas survey, it must be volatile – with a Henry's Law constant of at least 5×10^{-4} atm-m^3/mol and a vapor pressure of at least 1 mmHg (at 20 °C) (CCME, 1994). Notwithstanding, it should be recognized that, fine-grained materials of low gas permeability may serve as barriers to soil gas movement, resulting in the survey not detecting existing contamination because soil vapor is not able to reach the sampling point. High-water content materials (especially in perched water table zones) and sorptive organic-rich soils may also affect soil gas surveys in a similar manner – resulting in a possible reporting of 'false negatives'. Soil gas surveys may therefore not be used as an absolute indicator of site contamination – not even for suspected volatile organic contamination.

Overall, soil gas surveys are well-established methods for assessing the subsurface distribution of volatile organic contamination. They are particularly useful in the vicinity of storage tanks that are suspected to be leaking or in areas where some form of release or spill of fuels or organic solvents is suspected. In general, soil gas surveys are used to delineate the apparent extent of soil contamination and to identify locations for the collection of samples for more rigorous analysis conducted in an analytical laboratory. They can be a powerful screening technique, if used in a professionally credible and responsible manner. This is because, soil gas surveys can delineate source areas and track some contaminant plumes, allowing an investigator to place more accurately subsequent soil boring and monitoring well point locations.

Augering and borehole drilling Soil samples for laboratory analyses are collected using hand-held equipment (including hand augers, split spoon samplers, and ring samplers) or by advancing soil borings by the use of a drilling rig. In general, soil samples are collected from the apparent source areas as well as from suspected margins of the source areas in an effort to estimate the extents of contamination. Evidence of soil contamination in deeper borings that approach the water table suggest that groundwater is likely to be contaminated beneath and downgradient from the contaminant source area (Pratt, 1993). In any case, the volume of the release and the hydrogeological conditions are key factors in determining whether groundwater beneath the release site is indeed contaminated.

The locations of groundwater monitoring or sampling wells in relation to the potential on-site contaminant source areas and potential off-site receptors are critical in any data collection program. As a general rule-of-thumb, a *minimum* of three wells is necessary in order to determine the direction of groundwater flow at a project site; typically, one well is installed at the upgradient boundary of the site and at least two wells are positioned at locations downgradient from suspected source areas (Pratt, 1993). Realistically, additional wells located at the downgradient site boundary may be necessary to determine whether contaminated groundwater is flowing off-site. In fact, if it is determined that there possibly are seasonal changes in the groundwater flow directions, as occurs at some hydrogeologically complex sites, then even more wells may be necessary to account for the changes in contaminant migration directions.

2.3.2 Data Collection and Analysis Strategies

Traditionally, the characterization of soil contamination has been accomplished by taking several surface and subsurface soil samples, sealing them in sample containers, and shipping them for laboratory analyses. When the analytes of interest are volatile organic compounds (VOCs), the sample may be extracted with solvent upon arrival at the laboratory and the extract subsequently analyzed in a gas chromatograph (GC) or a gas chromatograph/mass spectrometer (GC/MS). Whereas these procedures tend to accommodate broad spectrum analysis and low levels of detection, they are also known to suffer severe limitations with respect to precision and accuracy. The failings of this approach are the result of both the heterogeneity of the soil medium and the ease with which VOCs vaporize and escape from the sample during the activities preceding the analysis. Consequently, soil sampling and analysis can grossly misrepresent VOC concentrations in soils by biasing it to the low end. At some sites, the underestimation can be so significant as to seriously affect the results of a site investigation effort, and ultimately the development of a site restoration program. It is noteworthy, however, that whereas individual soil samples may underestimate VOC levels, a suite of samples from across a site will reliably indicate if a contamination problem exists and will indeed characterize the nature of the plume adequately.

In general in areas where the contamination source is known, the sampling program should be targeted around that source. Normally sampling points should be located at regular distances along lines radiating from the contaminant source. Provisions should also be made in the site investigation to collect additional samples of small, isolated pockets of material which are visually suspect.

2.3.2.1   Soil vapor analysis as a general diagnostic tool for organic contamination

The presence of volatile contaminants or gas-producing materials can be determined by sampling the soil atmosphere within the ground. For example, organic vapor analyzer/ gas chromatograph (OVA/GC) or gas chromatograph/photoionization detector (GC/ PID) screening provides a relative measure of contamination by VOCs. Also, predictive models can be used to estimate the extent of gas migration from a suspected subsurface source. This information can subsequently be used to identify apparent 'hot spots', and to select soil samples for detailed chemical analyses. The on-site vapor analyses can also be helpful in selecting screened intervals for monitoring wells, and in the installation of a gas-monitoring well network (in conjunction with sampling in buildings in the area). Overall, soil vapor analysis offers a rapid and inexpensive alternative to soil analysis, by providing more representative data on contaminant concentrations.

On-site vapor screening of soil samples by surface probes and also during drilling can provide indicators of organic contamination. Using field GC equipment and hydraulically driven probes, soil vapors can be extracted and analyzed at various depths and several locations in a reasonably short period of time. Since results are obtained on a real-time basis, probe locations can be altered in the field to optimize positioning and complete characterization in a single mobilization. This reduces site characterization time and costs while assuring better coverage. In fact, an evaluation of data collected by several investigators from a variety of sites contaminated with VOCs has determined that the more up-to-date soil vapor techniques perform very well when properly applied.

Soil vapor data are indeed more representative of chemical contamination patterns because the probe draws a volume of vapor from the soil that lies radially distributed out from the sample location. In so doing, the probe blends vapors from a volume of soil and yields an average concentration for that sector of soil. In contrast, a single soil sample may be high or low depending on the probability that it incorporates a representative blend of particle sizes and mineral or organic content for the same sector of soil. Since VOCs generally concentrate on organic soil matters and clay, the soil sample needs to contain average levels of organics and clay to be representative.

In general, soil vapor analyses can be employed in the investigation of contaminated sites, to help determine the possibility for human exposures, the need for corrective measures, and the appropriate locations for monitoring wells and gas collection systems.

2.3.2.2   The investigation of sites with non-aqueous phase liquids

At most sites that are contaminated from spills and releases of organic solvents and/or fuels, the aqueous contaminant plume is usually detected first, whereas the so-called 'free product' or non-aqueous phase liquid (NAPL) – which actually serves as the source for long-term contaminant releases – tends to be elusive and rarely adequately defined.

During any site investigation activity, it is important to use some practical indicators of NAPL presence, in order to confirm or rule out the actual existence of such free product. Such indicators may include, but should not necessarily be limited to, information on known historical NAPL releases and/or usage, visible solvent product in subsurface samples (e.g. based on the appearance of saturated samples), high chemical concentrations in soil samples, very high OVA/PID readings, NAPL being detected in monitoring wells (as free product), and chemical concentrations detected in groundwater exceeding a specific compound's solubility limit (indicating possible presence of free product in well).

In fact, until the NAPL zone or area is identified and controlled or remediated, complete site characterization and subsequent cleanup of a project site is impossible. This is because solvent droplets or pools in the NAPL zone will serve as a long-term contaminant source that sustains any dissolved and/or vapor plume. In fact, concern about NAPLs exists because of their persistence in the subsurface, and their ability to contaminate large volumes of waters and soils due to the generally poor soil attenuating capability (Yong et al., 1992). A greater level of detail in the assessment of the transport and dissolution of NAPLs will aid in the development of cost-effective techniques for the control and cleanup of this type of contamination.

Soil gas analysis as a non-invasive method for NAPL site characterization Many NAPLs have high vapor pressures and will volatilize in the soil vadose zone to form a vapor plume around a NAPL source. VOCs dissolved in groundwater can also volatilize at the capillary fringe into soil gas. Soil gas surveys may therefore be used to help identify contaminated zones attributable to NAPL releases, which can then be subject to the additional investigations that are necessary to support site restoration decisions. Overall, it is believed that soil gas contamination will usually be dominated by volatilization and vapor transport from contaminant sources in the vadose zone rather than from groundwater, and that the upward transport of VOCs to the vadose zone from groundwater is probably limited to dissolved contaminants that are very near to the water table or to free product floating atop the water table.

2.3.2.3 The use of environmental immunochemical technologies

The advent of field-portable immunoassay methods has revolutionized many field and laboratory analyses used in the investigation of contaminated site problems. To date, environmental immunoassays have been developed and evaluated for a number of analytes (in a variety of matrices), including but not limited to major classes of pesticides, PCBs, PAHs, pentachlorophenol (PCP), the BTEX compounds (i.e. benzene, toluene, ethylbenzene, and xylenes), and some inorganic chemicals (Van Emon and Gerlach, 1995). All immunoassays rely on the interaction between an antibody (i.e. a protein that selectively recognizes and binds to a target analyte or group of related analytes) and a target analyte. Immunoassay test kits essentially package antibodies, reagents, standards, and substrates in field-transportable units that are ready for use. Sampling for immunoassay analysis will generally yield near-real-time data for potentially contaminated sites.

The use of immunoassays was the direct result of some recognized shortcomings of the GC in its application to certain chemical compounds. Although the immunoassay methods are not fully developed, it is apparent that the technique holds great promise for the future, and especially when used in conjunction with existing methods such as mass spectrometry. It is expected that field immunoassay analysis will gain increasing popularity and recognition, especially because of the highly portable equipment and minimal setup requirements. Furthermore, immunoassays can quickly and reliably map contamination present at contaminated sites, allowing subsequent sampling design to be more credible; this will then allow better use of in-lab analytical instrumentation and results.

Immunoassay can indeed be used in the laboratory as well as in the field. Its use in the pre-analysis of environmental samples prior to a GC analysis, for instance, can identify the need for dilution, thereby saving an expensive electron capture detector from contamination and downtime (Van Emon and Gerlach, 1995).

2.3.2.4   Optimizing the design of contaminated site monitoring systems

Groundwater monitoring programs are generally designed to investigate the possibility of spread of contamination from a contaminated site. But the adequacy of groundwater monitoring systems that consist simply of upgradient and downgradient monitoring wells is often questioned. This is because groundwater contaminants often migrate along preferred pathways that are the result of heterogeneities within stratigraphic units (Osiensky, 1995). Several monitoring wells must therefore be completed 'correctly' within individual heterogeneities to provide for the early detection of contaminant migration along these preferred flowpaths. Thus, the adequacy of a specific groundwater monitoring system is usually very much dependent on the number of monitoring wells.

A more innovative approach, involving the use of time series electrical potential field measurements in combination with groundwater monitoring wells, can greatly improve the effectiveness of groundwater monitoring systems required to detect the presence of site contamination (Osiensky, 1995). Typically, an electrical geophysical method is employed that involves the measurement of electrical current flow through the earth materials under investigation. Changes in the electric potential field over time due to the presence of groundwater contaminants can provide early detection of leaks from chemical disposal facilities or contaminated sites (Osiensky, 1995). In fact, geophysical methods have historically been used to help define potential pathways for contaminant migration. For example, surface electrical resistivity methods can be used to reduce the number of monitoring wells needed to define the migration of certain contaminants away from contaminated sites. Time series electrical potential measurements can be particularly useful for the early detection of releases from contaminated sites. Data from this can then be used to evaluate the potential for groundwater contamination, prior to drilling groundwater monitoring wells. The data can therefore help in the determination of the optimal locations for groundwater monitoring wells.

2.3.3   Optimization of the Overall Data Collection Process

A phased approach to environmental sampling generally encourages the early identification of key data requirements in the site investigation process. This ensures that the data collection effort is always directed at providing adequate information that meets the data quantity and quality requirements of the study. As a basic understanding of the site characteristics is achieved, subsequent data collection efforts focus on identifying and filling in any remaining data gaps from a previous phase of the site investigation. Any additionally acquired data should be such as to improve further the understanding of site characteristics, and also to consolidate information necessary to manage the contaminated site problem in an effective manner. In this way, the overall site investigation effort can be continually re-scoped to minimize the collection of unnecessary data and to maximize the quality of data acquired.

Overall, the data gathering process should provide a logical, objective, and quantitative balance between the time and resources available for collecting the data and the quality of data, based on the intended use of such data.

2.3.4 Evaluation of Environmental 'Control' Samples

Most environmental sampling and analysis procedures offer numerous opportunities for sample contamination from a variety of sources (Keith, 1988). To be able to address and account for possible errors arising from the non-site-related sources, quality control samples – called 'control' samples – are typically included in the sampling and analytical schemes. The 'control' samples are analytical quality control samples analyzed in the same manner as the environmental samples, and that are subsequently used in the measurement of any contamination that may have been introduced into a sample along its life cycle from the field (i.e. point of collection) to the laboratory (i.e. place of analysis).

Environmental 'control' samples can indeed be a very important reference datum for the evaluation of contaminated site sampling data. The analysis of environmental 'control' samples provides a way to determine if contamination has been introduced into a sample set either in the field while the samples were being collected and transported to the laboratory, or in the laboratory during sample preparation and analysis.

To prevent the inclusion of non-site-related constituents in the characterization of site contamination, the concentrations of the chemicals detected in 'control' samples must be compared with concentrations of the same chemicals detected in the environmental samples from the contaminated site. In general, 'control' samples containing common laboratory contaminants are evaluated differently from 'control' samples which contain chemicals that are not common laboratory contaminants. Thus, if the 'control' samples contain detectable levels of known common laboratory contaminants (e.g. acetone, 2-butanone [methyl ethyl ketone], methylene chloride, toluene, and the phthalate esters), then the environmental sample results may be considered as positive only if the concentrations in the sample exceed approximately ten times the maximum amount detected in any 'control' sample (DTSC, 1994; USEPA, 1989b, 1990). For 'control' samples containing detectable levels of one or more organic or inorganic chemicals that are not considered to be common laboratory contaminants, site sample results may be considered as positive only if the concentration of the chemical in the site sample exceeds approximately five times the maximum amount detected in any 'control' sample (DTSC, 1994; USEPA, 1989b, 1990). Invariably, environmental 'control' samples become an essential component of any site investigation program. This is because firm conclusions cannot be drawn from the site investigation unless adequate controls, have been included as part of the sampling and analytical protocols (Keith, 1988).

2.3.5   Treatment of Censored Data in Environmental Samples

Environmental data sets may contain observations which are below the instrument or method detection limit, or its corresponding quantitation limit; such data are often referred to as 'censored data' (or 'non-detects' [NDs]). Invariably, all laboratory analytical techniques have detection and quantitation limits below which only 'less than' values may be reported; the reporting of such values provides a degree of quantification for the censored data. In fact, it is customary to assign non-zero values to all environmental sampling data reported as NDs. This is important because, even at or near their detection limits, certain chemical constituents may be of considerable importance in the characterization of a contaminated site. However, uncertainty about the actual values below the detection or quantitation limit can bias or preclude subsequent statistical analyses. Censored data do indeed create great uncertainties in the data analysis required of the site characterization process; such data should therefore be handled in an appropriate manner, as elaborated below.

Oftentimes, in a given set of environmental samples, certain chemicals will be reliably quantified in some, but not all, of the samples. This situation may reflect the fact that either the chemical is truly absent at this location at the time the sample was collected, or the chemical is present but at a concentration below the quantitation limits of the analytical method that was employed in the sample analysis. In such situations, a decision has to be made as to how to treat such NDs and associated 'proxy' concentrations. The appropriate procedure, which depends on the pattern of detection of the chemical across the entire site, may consist of the following determinations (HRI, 1995):

- If a chemical is rarely detected in any medium, and there is little to no reason to expect the chemical to be a site contaminant, then it may be appropriate to exclude it from further analysis. This is particularly true if the chemical is a common laboratory contaminant.
- If a chemical is rarely detected in a specific medium, and there is no reason to expect the chemical to be a significant site contaminant, then it may be appropriate to exclude it from further analysis of that particular medium.
- If the pattern of a chemical's concentration in samples suggests that it was confined to well-defined 'hot spots' at the time of sampling, then the potential for contaminant migration should be considered. Consequently, it will be important to include such a chemical in the analysis of both the source and receiving media.

Derivation and use of 'proxy' concentrations   'Proxy' concentrations are usually employed when a chemical is not detected in a specific medium. A variety of approaches are offered in the literature for deriving and using proxy values in environmental data analyses, including the following (HRI, 1995; USEPA, 1989a):

- *Set the sample concentration to zero.* This involves very strong assumptions, rarely justified, that the chemical is not present in the environmental samples.

- *Drop the sample with the non-detect for the particular chemical from further analysis.* This will have the same effect on the data analysis as assigning a concentration that is the average of concentrations found in samples where the chemical was detected.
- *Set the proxy sample concentration to the sample quantitation limit (SQL).* For NDs, setting the sample concentration to a proxy concentration equal to the SQL (which is a quantifiable number used in practice to define the analytical detection limit) makes the fewest assumptions and tends to be conservative, since the SQL represents an upper-bound on the concentration of a ND. This approach recognizes that the true distribution of concentrations represented by the NDs is unknown.
- *Set the proxy sample concentration to one-half the SQL.* For NDs, setting the sample concentration to a proxy concentration equal to one-half the SQL assumes that, regardless of the distribution of concentrations above the SQL, the distribution of concentrations below the SQL is symmetrical.

The common practice involves specifying the sample-specific quantitation limit for the chemical.

Notwithstanding the above procedures of using 'proxy' concentrations, re-sampling should always be viewed as the preferred approach to resolving uncertainties that surround ND results from environmental samples. Under such circumstances, if the initially reported data represent a problem in sample collection or analytical methods rather than a true failure to detect, then the identified problem may be rectified (e.g. by the use of more sensitive analytical protocols) before critical decisions are made based on the earlier results.

A method of approach for the statistical evaluation of NDs   The favored approach in the calculation of the applicable statistical values during the evaluation of data containing NDs involves the use of a value of one-half of the SQL. This approach assumes that the samples are equally likely to have any value between the detection limit and zero, and can be described by a normal distribution. However, when the sample values above the ND level are lognormally distributed, it generally may be assumed that the ND values are also lognormally distributed. The best estimate of the ND values for a lognormally distributed data set is the reported SQL divided by the square root of two (i.e. SQL/1.414) (CDHS, 1990; USEPA, 1989a).

In general, during the analysis of environmental sampling data that contain some NDs, a fraction of the SQL is usually assumed (as a proxy or estimated concentration) for non-detectable levels, instead of assuming a value of zero or neglecting such values. This procedure is used, provided there is at least one detected value from the analytical results, and/or if there is reason to believe that the chemical is possibly present in the sample at a concentration below the SQL. This approach conservatively assumes that some level of the chemical could be present (even though a ND has been recorded) and arbitrarily sets that level at the appropriate percentage of the SQL (i.e. SQL/2 if data set is assumed to be normally distributed or SQL/1.414 for a lognormally distributed data set). In fact, in some situations the SQL value itself may be used if there is strong reason to believe that the chemical concentration is closer to this value, rather than to a fraction of the SQL. Where it is apparent that serious biases could result, more sophisticated analytical and evaluation methods may be warranted.

2.3.6   General Approach to the Statistical Analysis of Environmental
Sampling Data

Several of the available statistical methods and procedures finding widespread use in contaminated site characterization programs can be found in the literature on statistics (e.g. Cressie, 1994; Freund and Walpole, 1987; Gibbons, 1994; Gilbert, 1987; Hipel, 1988; Miller and Freund, 1985; Sharp, 1979; Wonnacott and Wonnacott, 1972; Zirschy and Harris, 1986). Some commonly used methods of approach that find general application in the evaluation of environmental data are briefly discussed below.

Over the years, extensive technical literature has been developed regarding the *best* probability distribution to utilize in different scientific applications. Of the many statistical distributions available, the Gaussian (or normal) distribution has been widely utilized to describe environmental data. However, there is considerable support for the use of the lognormal distribution in describing environmental data; the use of lognormal statistics for the data set X_1, X_2, X_3, . . . , X_n requires that the logarithmic transform of these data (i.e. $\ln[X_1]$, $\ln[X_2]$, $\ln[X_3]$, . . . , $\ln[X_n]$) can be expected to be normally distributed. Consequently, chemical concentration data in the environment have been described by the lognormal distribution, rather than by a normal distribution (Gilbert, 1987; Leidel and Busch, 1985; Rappaport and Selvin, 1987). In fact, the use of a normal distribution (whose central tendency is measured by the arithmetic mean) to describe environmental contaminant distribution, rather than lognormal statistics (whose central tendency is defined by the geometric mean) will often result in significant overestimation, and may be overly conservative.

More sophisticated methods, such as geostatistical techniques that account for spatial variations in concentrations, may also be employed for estimating the average concentrations at a site (e.g. USEPA, 1988a; Zirschy and Harris, 1986). In particular, a technique called block kriging is frequently used to estimate soil chemical concentrations in sections of contaminated sites in which only sparse sampling data exist. In this case, the site is divided into blocks (or grids), and concentrations are determined within blocks by using interpolation procedures that incorporate sampling data in the vicinity of the block. The sampling data are weighted in proportion to the distance of the sampling location from the block.

The statistical evaluation of environmental data   Statistical procedures used for the evaluation of environmental data can significantly affect the conclusions of a given site characterization program. Consider, for instance, the use of a normal distribution (whose central tendency is measured by the arithmetic mean) to describe environmental contaminant distribution, rather than lognormal statistics (whose central tendency is defined by the geometric mean); the former will often result in significant overestimation of contamination levels. Appropriate statistical methods should therefore be utilized in the evaluation of environmental sampling data (e.g. in relation to the choice of proper averaging techniques). Furthermore, contamination levels and exposures may have temporal variations and the dynamic nature of such parameters should, insofar as possible, be incorporated in the evaluation of the environmental data.

In general, statistical procedures used in the evaluation of environmental data should reflect the character of the underlying distribution of the data set. The appropriateness of any distribution assumed or used for a given data set should preferably be checked prior to its application; this can be accomplished by using some goodness-of-fit methods (see, e.g., Cressie, 1994; Freund and Walpole, 1987; Gilbert, 1987; Miller and Freund, 1985; Sharp, 1979; Wonnacott and Wonnacott, 1972). The choice of statistical parameters for site characterization programs is also critical to the corrective action decisions about a contaminated site problem.

2.3.6.1 Selection of statistical averaging techniques

The selection of appropriate methods of approach to averaging a set of environmental sampling data can have profound effects on the resulting concentration, especially for data sets of sampling results that are not normally distributed. Consequently, reasonable discretion should be exercised in the selection of an averaging technique during the analysis of environmental sampling data. For example, when dealing with lognormally distributed data, geometric means are often used as a measure of central tendency, to ensure that a few very high values do not exert excessive influence on the characterization of the distribution. If, however, high concentrations do indeed represent 'hot spots' in a spatial distribution, then using the geometric mean would inappropriately discount the contribution of these high concentrations present at a contaminated site. This is particularly true if the spatial pattern indicates that areas of high concentration are in close proximity to compliance boundaries or points of exposure to populations potentially at risk.

The geometric mean has indeed been extensively used as an averaging parameter in the past. Its principal advantage is in minimizing the effects of 'outlier' values (i.e. a few values that are much higher or lower than the general range of sample values). Its corresponding disadvantage is that discounting these values may be inappropriate, when they represent true variations in concentrations from one part of a site to another (such as a 'hot spot' vs. a 'cold spot'). As a measure of central tendency, the geometric mean is most appropriate if sample data are lognormally distributed, without an obvious spatial pattern.

The arithmetic mean – commonly used when referring to an 'average' – is more sensitive to a small number of extreme values or a single 'outlier' compared to the geometric mean. Its corresponding advantage is that true high concentrations will not be inappropriately discounted. With limited sampling data, however, this may not provide a conservative enough estimate of site contamination.

In fact, none of the above measures, in themselves, may be appropriate in the face of limited and variable sampling data. Current applications tend to favor the use of an upper confidence limit on the average concentration. In situations where there is a discernible spatial pattern to contaminant concentration data, standard approaches to data aggregation and analysis may usually be inadequate or even inappropriate.

Illustration of the effects of statistical averaging techniques on concentration predictions To demonstrate the possible effects of the choice of statistical distributions and/or averaging techniques on the analysis of environmental data, consider a case involving the estimation of the mean, standard deviation, and confidence limits of monthly groundwater sampling data from a contaminated site. In order to compare the selected statistical parameters based on the assumption that these data are normally distributed *versus* an alternative assumption that the data are lognormally distributed, the several statistical manipulations enumerated below are carried out on the 'raw' and log-transformed data for the concentrations of benzene in the groundwater samples shown in Table 2.2.

Table 2.2   Environmental sampling data used to illustrate the effects of averaging techniques on concentration predictions

| Sampling event | Concentration of benzene in water (μg/L) | |
	Original 'raw' data, X	Log-transformed data, $Y = \ln(X)$
1	0.049	-3.016
2	0.056	-2.882
3	0.085	-2.465
4	1.200	0.182
5	0.810	-0.211
6	0.056	-2.882
7	0.049	-3.016
8	0.048	-3.037
9	0.062	-2.781
10	0.039	-3.244
11	0.045	-3.101
12	0.056	-2.882

1. Calculate the following statistical parameters for the 'raw' data: mean, standard deviation, and 95% confidence limits (see standard statistics textbooks for details of procedures involved). The arithmetic mean, standard deviation, and 95% confidence limits (95% CL) for a set of n values are defined, respectively, as follows:

$$X_m = \frac{\sum_{i=1}^{n} X_i}{n}, \quad s = \left[\frac{\sum_{i=1}^{n}(X_i - X_m)^2}{(n-1)} \right]^{0.5}, \quad \text{and } CL = X_m + \frac{ts}{n^{0.5}}$$

where t is the value of the student's t distribution (refer to standard statistical texts) for the desired confidence level and degrees of freedom, $(n-1)$, and s is an estimate of the standard deviation from the mean (X_m). Thus,

$X_m = 0.213 \, \mu$g/L

$SD_x = 0.379 \, \mu$g/L

$CL_x = 0.213 \pm 0.241 \, \mu$g/L (i.e. $-0.028 \leqslant CI_x \leqslant 0.454$) and $UCL_x = 0.454 \, \mu$g/L

where:

X_m = arithmetic mean of 'raw' data

SD_x = standard deviation of 'raw' data

CI_x = 95% confidence interval (95% CI) of 'raw' data

UCL_x = 95% upper confidence level (95% UCL) of 'raw' data

Note that, the development of a 95% confidence limit for the untransformed data gives a confidence interval of $0.213 \pm 0.109t = 0.213 \pm 0.241$ (where $t = 2.20$, obtained from the student's t distribution for $(n-1) = 12 - 1 = 11$ degrees of freedom), indicating a non-zero probability of a negative concentration value.

2. Calculate the following statistical parameters for the log-transformed data: mean, standard deviation, and 95% confidence limits (see standard statistics textbooks for details of procedures involved). The geometric mean, standard deviation, and 95% confidence limits (95% CL) for a set of n values are defined, respectively, as follows:

$$X_{gm} = \text{antilog} \left\{ \frac{\sum_{i=1}^{n} \log X_i}{n} \right\}, \quad s = \left[\frac{\sum_{i=1}^{n} (X_i - X_{gm})^2}{(n-1)} \right]^{0.5}, \quad \text{and CL} = X_{gm} + \frac{ts}{n^{0.5}}$$

where t is the value of the student's t distribution (refer to standard statistical texts) for the desired confidence level and degrees of freedom, $(n-1)$, and s is an estimate of the standard deviation of the mean (X_{gm}). Thus,

$Y_{\text{a-mean}} = -2.445$

$SD_y = 1.154$

$CL_y = -2.445 \pm 0.733$ (i.e. a confidence interval from -3.178 to -1.712)

where:

$Y_{\text{a-mean}} = $ arithmetic mean of log-transformed data

$SD_y = $ standard deviation of log-transformed data

$CI_y = $ 95% confidence interval (95% CI) of log-transformed data

The development of a 95% confidence limit for the log-transformed data gives a confidence interval of $-2.445 \pm 0.348t = -2.445 \pm 0.765$ (where $t = 2.20$, obtained from the student's t distribution for $n = 12 - 1 = 11$ degrees of freedom).

Transforming the average of the Y values back into arithmetic values yields a geometric mean value, $X_{gm} = e^{-2.445} = 0.087$. Furthermore, transforming the confidence limits of the log-transformed values back into the arithmetic realm yields a 95% confidence interval of 0.042 to $0.180 \, \mu g/L$, consisting of positive concentration values only. Thus,

$X_{gm} = 0.087$

$0.042 \leqslant CI_x \leqslant 0.180 \, \mu g/L$

$UCL_x = 0.180 \, \mu g/L$

where:

$X_{gm} = $ geometric mean for the 'raw' data

$CI_x = $ 95% confidence interval (95% CI) for the 'raw' data (assuming lognormal distribution)

$UCL_x = $ 95% upper confidence level (95% UCL) for the 'raw' data (assuming lognormal distribution)

It is apparent that the arithmetic mean, $X_m = 0.213 \, \mu g/L$, is substantially larger than the geometric mean of $X_{gm} = 0.087 \, \mu g/L$. The reason for this is that two large sample values in the data set (i.e. sampling events numbers 4 and 5 in Table 2.2) tend to strongly bias the arithmetic mean; the logarithmic transform acts to suppress the extreme values. A similar observation can be made for the 95% upper confidence level (UCL) of the normally and lognormally distributed data sets. In general, however, the 95% UCL is a preferred statistical parameter to use in the evaluation of environmental data rather than the mean values, irrespective of the type of underlying distribution.

The results from this example analysis illustrate the potential effects that could result from the choice of one distribution type over another, and also the implications of selecting specific statistical parameters in the evaluation of environmental sampling data. In general, the use of arithmetic or geometric mean values for estimating average concentrations would tend to bias the concentration estimates.

2.3.7   Using 'Control' Sites to Establish Background Thresholds

Control sites, considered not to have been impacted by a contaminated site, are important to understanding the significance of environmental sampling and monitoring data. Locations selected as control sites should generally have similar characteristics (i.e. identical in their physical and environmental settings) as the potentially contaminated area under investigation. Background samples collected at control sites are typically used to demonstrate whether or not a site is truly contaminated. The background sampling results allow a technically valid scientific comparison to be made between environmental samples (suspected of containing site contaminants) and control site samples (possibly containing only naturally-low or anthropogenic levels of the same chemicals).

There are two types of control sites – local and area – whose differentiation is based primarily on the closeness of the control site to the environmental sampling or project site. Local control sites are usually adjacent or very near the potentially impacted project sites, whereas an area control site is in the same general area or region as the project site, *but not adjacent to it*. In selecting and working with either type of control sites, the following principles and factors should be taken into consideration (Keith, 1991; CCME, 1993):

- Control sites generally should be upwind, upstream, and/or upgradient from the environmental sampling site.
- When possible, control site samples should be taken first to avoid possible cross-contamination from the environmental sampling site.
- Travel between control sites and environmental sampling areas should be minimized because of potential cross-contamination caused by humans, equipment, and/or vehicles.

In general, local control sites are preferable to area control sites because they are physically closer. However, when a suitable local control site cannot be found, an area control site will generally provide for the requisite background sampling information.

Sampling information from the control sites is subsequently used to establish background thresholds. The background threshold is meant to give an indication of the level of contamination in the environment that may not necessarily be attributed to the potentially contaminated site under investigation. This serves to provide a reference 'point-of-departure' that can be used to determine the magnitude of contamination in other environmental samples obtained from a contaminated site. Ideally, background samples (or control site samples) are preferably collected near the time and place of the environmental samples of interest.

2.3.7.1 Background sampling requirements

Background (or control site) samples are typically collected and evaluated to determine the possibility of a potentially contaminated site contributing to off-site contamination levels in the vicinity of the case site. The background samples would not have been significantly influenced by contamination from the project site. However, these samples are obtained from an environmental matrix that has similar basic characteristics as the matrix at the project site, in order to provide a justifiable basis for comparison.

To satisfy acceptable criteria required of background samples, the following requirements should be carefully incorporated in to the design of background sampling programs:

- *Significance of matrix effects on environmental contaminant levels.* In a number of investigations, the analysis of data from different soil types in the same background area revealed levels of select inorganic constituents that were over twice as high in silt/clay as in sand (e.g. LaGoy and Schulz, 1993). Thus, it is important to give adequate consideration to the effects of natural variations in soil composition when one is designing a field sampling program. In fact, unless background samples are collected and analyzed under the same conditions as the environmental test samples, the presence and/or levels of the analytes of interest and the effects of the matrix on their analysis cannot be known or estimated with any acceptable degree of certainty. Therefore, background samples of each significantly different matrix must always be collected when different types of matrices are involved, such as various types of water, sediments, and soils in or near a sampling site area.
- *Number of background samples.* A minimum of three background samples per medium will usually be collected, although more may be desired especially in complex environmental settings. In general, if the natural variability of a particular constituent present at a site is relatively large, the sampling plan should reflect this site-specific characteristic.
- *Background sampling locations.* In typical sampling programs, background air samples would consist of upwind air samples and, perhaps, different height samples; background soil samples would be collected near a site in areas upwind and upslope of the site; background groundwater samples generally come from upgradient well locations, in relation to groundwater flow direction(s) at the site; and background surface water and sediment samples may be collected under both high and low flow conditions at upstream locations, and insofar as practicable, sample collection from nearby lakes and wetlands should comprise of shallow and deep samples (when sufficient water depth allows), to account for such differences potentially resulting from stratification or incomplete mixing.

More detailed background sampling considerations and strategies for the various environmental media of general interest can be found in the literature elsewhere (e.g. Keith, 1991; Lesage and Jackson, 1992; USEPA, 1988b, 1989b). In general, background sampling is conducted to distinguish site-related contamination from naturally-occurring or other non-site-related levels of select constituents. Anthropogenic levels (which are concentrations of chemicals that are present in the environment due to human-made, non-site sources, such as industry and automobiles), rather than naturally-occurring levels, are preferably used as a basis for evaluating background sampling data.

2.4 EVALUATING THE SIGNIFICANCE OF SITE CONTAMINATION

Decisions about the significance of observed site contamination should generally be formulated in the context of the site-specific scenarios associated with a project site. The mobility of the site contaminants, the location of populations potentially at risk, the size and shape of contaminant plumes, the presence or absence of free products or NAPLs, background threshold concentrations, and the intended use of the site are all important factors to consider in assessing the significance of the contamination (Pratt, 1993). These factors can be integrated in to a coherent structure by developing a conceptual model for the site that shows the inter-relationship between the contaminant source locations, transport media, and potential exposure point locations or compliance boundaries. Exposure scenarios can subsequently be developed based on the site-specific conditions and the anticipated land-uses; the risk posed to the critical receptors is then determined and the cleanup goals defined in terms of an 'acceptable' risk level.

2.5 REFERENCES

BSI (British Standards Institution) (1988). Draft for Development, DD175: 1988 Code of Practice for the Identification of Potentially Contaminated Land and its Investigation. BSI, London, UK.

Cairney, T. (ed.) (1993). *Contaminated Land (Problems and Solutions)*. Blackie Academic & Professional, Glasgow/Chapman & Hall, London/Lewis Publishers, Boca Raton, Florida.

CCME (Canadian Council of Ministers of the Environment) (1993). *Guidance Manual on Sampling, Analysis, and Data Management for Contaminated Sites*. Volume I: Main Report (Report CCME EPC-NCS62E), and Volume II: Analytical Method Summaries (Report CCME EPC-NCS66E). The National Contaminated Sites Remediation Program, Winnipeg, Manitoba, December 1993.

CCME (Canadian Council of Ministers of the Environment) (1994). *Subsurface Assessment Handbook for Contaminated Sites*. Canadian Council of Ministers of the Environment (CCME), The National Contaminated Sites Remediation Program (NCSRP), Report No. CCME-EPC-NCSRP-48E (March 1994), Ottawa, Ontario, Canada.

CDHS (California Department of Health Services) (1990). *Scientific and Technical Standards for Hazardous Waste Sites*. Prepared by the California Department of Health Services, Toxic Substances Control Program, Technical Services Branch, Sacremento, California.

Cressie, N.A. 1994. *Statistics for Spatial Data*, revised edn. John Wiley & Sons, New York.

DTSC (Department of Toxic Substances Control) (1994). *Preliminary Endangerment Assessment Guidance Manual* (A guidance manual for evaluating hazardous substance release sites). California Environmental Protection Agency, DTSC, Sacramento, California.

Freund, J.E. and R.E. Walpole (eds) (1987). *Mathematical Statistics*. Prentice-Hall, Englewood Cliffs, New Jersey.

Gibbons, R.D. (1994). *Statistical Methods for Groundwater Monitoring*. John Wiley & Sons, New York.

Gilbert, R.O. (1987). *Statistical Methods for Environmental Pollution Monitoring.* Van Nostrand Reinhold, New York.

Hipel, K.W. (1988). Nonparametric approaches to environmental impact assessment. *Water Resources Bull.* **24**(3), 487–491.

HRI (Hampshire Research Institute) (1995). *Risk*Assistant for Windows.* The Hampshire Research Institute, Inc., Alexandria, Virginia.

Keith, L.H. (1988). *Principles of Environmental Sampling.* American Chemical Society (ACS), Washington, DC.

Keith, L.H. (1991). *Environmental Sampling and Analysis – A Practical Guide.* Lewis Publishers, Boca Raton, Florida.

LaGoy, P.K. and C.O. Schulz (1993). Background sampling: an example of the need for reasonableness in risk assessment. *Risk Anal.* **13**(5), 483–484.

Leidel, N. and K.A. Busch (1985). Statistical design and data analysis requirements. In: *Patty's Industrial Hygiene and Toxicology*, Vol. IIIa, 2nd edn, John Wiley & Sons, New York.

Lesage, S. and R.E. Jackson (eds) (1992). *Groundwater Contamination and Analysis at Hazardous Waste Sites.* Marcel Dekker, Inc., New York.

Miller, I. and J.E. Freund (1985). *Probability and Statistics for Engineers*, 3rd edn. Prentice-Hall, Englewood Cliffs, New Jersey.

Osiensky, J.L. (1995). Time series electrical potential field measurements for early detection of groundwater contamination. *J. Environ. Sci. Health* **A30**(7), 1601–1626.

Pratt, M. (ed.) (1993). *Remedial Processes for Contaminated Land.* Institution of Chemical Engineers, Warwickshire, UK.

Rappaport, S.M. and J. Selvin (1987). A method for evaluating the mean exposure from a lognormal distribution. *J. Amer. Ind. Hyg. Assoc.* **48**, 374–379.

Sharp, V.F. (1979). *Statistics for the Social Sciences.* Little, Brown & Co., Boston, Massachusetts.

Telford, W.M., L.P. Geldart, R.E. Sheriff, and D.A. Keys (1976). *Applied Geophysics.* Cambridge University Press, New York.

USEPA (US Environmental Protection Agency) (1985). *Characterization of Hazardous Waste Sites – A Methods Manual, Volume 1: Site Investigations.* Environmental Monitoring Systems Laboratory, Las Vegas, Nevada. EPA/600/4-84/075 (April 1985).

USEPA (US Environmental Protection Agency) (1988a). *GEO-EAS (Geostatistical Environmental Assessment Software) User's Guide.* Environmental Monitoring Systems Laboratory, Office of R&D, Las Vegas, Nevada. EPA/600/4-88/033a.

USEPA (US Environmental Protection Agency) (1988b). *Guidance for Conducting Remedial Investigations and Feasibility Studies Under CERCLA.* EPA/540/G-89/004. OSWER Directive 9355.3-01, Office of Emergency and Remedial Response, Washington, DC.

USEPA (US Environmental Protection Agency) (1989a). *Risk Assessment Guidance for Superfund. Volume I – Human Health Evaluation Manual (Part A).* EPA/540/1-89/002. Office of Emergency and Remedial Response, Washington, DC.

USEPA (US Environmental Protection Agency) (1989b). *Soil Sampling Quality Assurance User's Guide*, 2nd edn. EPA/600/8-89/046, Experimental Monitoring Support Laboratory (EMSL), ORD, US EPA, Las Vegas, Nevada.

USEPA (US Environmental Protection Agency) (1990). *Guidance for Data Usability in Risk Assessment, Interim Final.* Office of Emergency and Remedial Response. EPA/540/G-90/008, Washington, DC.

Van Emon, J.M. and C.L. Gerlach (1995). A Status Report on Field-Portable Immunoassay. *J. Environ. Sci. Technol.* **29**(7), 312A–317A.

Wonnacott, T.H. and R.J. Wonnacott (1972). *Introductory Statistics*, 2nd edn. John Wiley & Sons, New York.

Young, R.N., A.M.O. Mohamed, and B.P. Warkentin (1992). *Principles of Contaminant Transport in Soils. Developments in Geotechnical Engineering*, Vol. 73. Elsevier Scientific Publishers BV, Amsterdam, The Netherlands.

Zirschy, J.H. and D.J. Harris (1986). Geostatistical analysis of hazardous waste site data. *ASCE J. Environ. Engnr.* **112**(4).

Chapter Three

Contaminant fate and transport in the environment

Contaminants released into the environment are controlled by a complex set of processes consisting of transport, transformation, degradation and decay, cross-media transfers, and/or biological uptake and bioaccumulation. Environmental fate and transport analyses offer a way to assess the movement of chemicals between environmental compartments, and the prediction of the long-term fate of such chemicals in the environment.

In general, as pollutants are released into various environmental media, several factors contribute to their migration from one environmental matrix into another, or their phase changes from one physical state into another. The relevant phenomena involved in the fate and transport of environmental contaminants, together with the important factors affecting the processes involved, are annotated in this chapter.

3.1 THE CROSS-MEDIA TRANSFER OF CONTAMINANTS BETWEEN ENVIRONMENTAL COMPARTMENTS

Chemicals present in one environmental matrix may be affected by several complex processes and phenomena, facilitating transfers into other media. The potential for cross-media transfers of pollutants from the soil medium into other media is particularly significant. In fact, the movement of contaminants through soils is generally very complex, with some pollutants moving rapidly while others move rather slowly. In general, the affinity that contaminants have for soil affects their mobility by retarding transport. For instance, hydrophobic or cationic contaminants that are migrating in solution are subject to retardation effects. The hydrophobicity of a contaminant can greatly affect its fate, which explains some of the different rates of contaminant migration that occur in the subsurface environment. On the other hand, chemical constituents having a moderate to high degree of mobility can leach from soils into groundwater; volatile constituents may contribute to subsurface gas in the vadose zone, and also possible releases into the atmosphere. Conversely, it is possible for cross-media transport of constituents from other media into soils to take place; for example, chemical constituents may be transported into the soil matrix via deposition of suspended particulates from the atmosphere, and also through releases of subsurface gas.

Contaminated soil is indeed the main source of chemical repository for most environmental pollutants. However, the soil medium is by no means an inert repository, since there is an active interchange of chemicals between soils and water, air, and biota

(Asante-Duah, 1993; Brooks et al., 1995), with the main driving forces in contaminant transport in soils being advective and diffusive in nature (Yong et al., 1992). Consequently, oftentimes, soils become the principal focus of attention in the investigation of contaminated sites. This is reasonable because soils at such sites not only serve as a medium of exposure to potential receptors, but also serve as a long-term reservoir for contaminants to be released into other media.

3.1.1   Phase Distribution and Cross-media Transfers of Environmental Contaminants

Contamination occurring at contaminated sites may exist in different physical states, and may generally be present in a variety of environmental matrices, in particular soils and groundwater. For example, when fluids that are immiscible with water, called non-aqueous phase liquids (NAPLs), are released at a site and then enter the subsurface environment, they tend to exist as distinct fluids that flow separately from the water phase. Fluids less dense than water, or light NAPLs (LNAPLs), migrate downward through the vadose (unsaturated) zone, but tend to form lenses that 'float' on top of an aquifer upon reaching the water table. Typical examples of LNAPLs include hydrocarbon fuels such as gasoline, heating oil, kerosene, jet fuel, and aviation gas. Such LNAPLs will pool and spread as a floating free product layer atop the water table if they are released to the subsurface in sufficient quantities. Denser-than-water NAPLs, or dense NAPLs (DNAPLs), will tend to 'sink' into the aquifer. Typical examples of DNAPLs include chlorinated hydrocarbons (e.g. trichloroethylene [TCE], tetrachloro-ethene [PERC], chlorophenols, chlorobenzenes, and PCBs), coal tar wastes, creosote-based wood-treating oils, some pesticides, etc. Such DNAPLs can pass across the water table and may be found at greater depths within the saturated zone of a groundwater aquifer.

Contact with groundwater or infiltrating recharge water causes some of the chemical constituents of a NAPL to dissolve, resulting in aquifer contamination, further to any trail of contamination left in the soil matrix. In fact, NAPLs may be present at several contaminated sites globally. However, due to the numerous variables influencing NAPL transport and fate in the subsurface, they are likely to go undetected and yet they are likely to be a significant limiting factor in site restoration decisions (Charbeneau et al., 1992). This is especially true for DNAPLs, such as chlorinated solvents.

In general, contamination present at contaminated sites may typically show up in the following phases:

- Adsorbed contamination (on to solid phase matter or soils).
- Vapor phase contamination (present in the vadose zone due to volatilization into soil gas).
- Dissolved contamination (in water, present in both the unsaturated and saturated soil zones).
- Free product (as residual and mobile immiscible fluids, e.g. as LNAPL floating on the surface of the water table, or as DNAPL that sinks deep into the groundwater zone, or as NAPL persisting in the soil pore spaces).

This contamination system that exists in the soils, water, vapor phase, or the NAPLs tends towards a state of equilibrium, such that the chemical potential or fugacity is equal in all the phases that co-exist in the environment. Changes in equilibrium between the phases can be occurring on a continuing basis as a result of several extraneous factors that affect the concentration gradients.

3.2   IMPORTANT FATE AND TRANSPORT PROPERTIES, PROCESSES, AND PARAMETERS

As pollutants are released into various environmental media, several factors contribute to their migration and transport. Examination of a contaminant's physical and chemical properties can often allow an estimation of its degree of environmental partitioning, migration and/or attenuation. Several important physical and chemical properties, processes, and parameters affecting the environmental fate and/or cross-media transfers of environmental contaminants are briefly annotated below, with further detailed discussions offered elsewhere in the literature (e.g. Evans, 1989; Lyman et al., 1990; Samiullah, 1990; Swann and Eschenroeder, 1983; USEPA, 1989; Yong et al., 1992).

3.2.1   Physical State

At a typical contaminated site, contaminants may exist in all three major physical states (viz.: solids adsorbed on to soils, free product or dissolved contaminants, and vapor phase in the soil vadose zones). Contaminants in the solid phase are generally less susceptible to release and migration than fluids. However, processes such as leaching, erosion and/or runoff, and physical transport of chemical constituents can act as significant release mechanisms, irrespective of the physical state of a contaminant.

3.2.2   Water Solubility

The solubility of a chemical in water is the maximum amount of the chemical that will dissolve in pure water at a specified temperature. Solubility is an important factor affecting a chemical constituent's release and subsequent migration and fate in the surface water and groundwater environments. In fact, among the various parameters affecting the fate and transport of organic chemicals in the environment, water solubility is one of the most important, especially with regards to hydrophilic compounds.

Typically, solubility affects mobility, leachability, availability for biodegradation, and the ultimate fate of a given constituent. In general, highly soluble chemicals are easily and quickly distributed by the hydrologic system. Such chemicals tend to have relatively low adsorption coefficients for soils and sediments, and also relatively low bioconcentration factors in aquatic biota. Furthermore, they tend to be more readily biodegradable. Substances which are more soluble are more likely to desorb from soils and less likely to volatilize from water.

3.2.3 Diffusion

Diffusive processes create mass spreading due to molecular diffusion, in response to concentration gradients. The higher the diffusivity, the more likely a chemical is to move in response to concentration gradients. Thus, diffusion coefficients are used to describe the movement of a molecule in a liquid or gas medium as a result of differences in concentration. The diffusion coefficient can also be used to calculate the dispersive component of chemical transport.

3.2.4 Dispersion

Dispersive processes create mass mixing due to system heterogeneities (e.g. velocity variations). Consequently, for example, as a pulse of contaminant plume migrates through a soil matrix, the peaks in concentration are decreased by spreading. Dispersion is an important attenuation mechanism that results in the dilution of a contaminant; the degree of spreading or dilution is proportional to the size of the dispersion coefficients.

3.2.5 Volatilization

Volatilization is the process by which a chemical compound evaporates from one environmental compartment into the vapor phase. The volatilization of chemicals is a very important mass-transfer process. The transfer process from the source (e.g. water-body, sediments, or soils) to the atmosphere is dependent on the physical and chemical properties of the compound in question, the presence of other pollutants, the physical properties of the source media, and the atmospheric conditions.

Knowledge of volatilization rates is important in the determination of the amount of chemicals entering the atmosphere and the change of pollutant concentrations in the source media. Volatility is therefore considered a very important parameter for chemical hazard assessments. Several important measures of a chemical's volatility or volatilization rate are enumerated below.

- *Henry's Law constant.* Henry's Law constant (H) provides a measure of the extent of chemical partitioning between air and water at equilibrium. It indicates the relative tendency of a constituent to volatilize from aqueous solution into the atmosphere, based on the competition between its vapor pressure and water solubility. This parameter is important to determining the potential for cross-media transport into air.

  Contaminants with low Henry's Law constant values will tend to favor the aqueous phase and will therefore volatilize into the atmosphere more slowly than constituents with high values. As a general guideline: H values in the range of 10^{-7} to 10^{-5} (atm-m^3/mol) represent low volatilization; H between 10^{-5} and 10^{-3} (atm-m^3/mol) means volatilization is not rapid but possibly significant; and $H > 10^{-3}$ (atm-m^3/mol) implies volatilization is rapid. The variation in H between chemicals is indeed extensive.

- *Vapor pressure.* Vapor pressure is the pressure exerted by a chemical vapor in equilibrium with its solid or liquid form at any given temperature. It is a relative measure of the volatility of a chemical in its pure state, and is an important determinant of the rate of volatilization. The vapor pressure of a chemical can be used to calculate the rate of volatilization of a pure substance from a surface, or to estimate a Henry's Law constant for chemicals with low water solubility.

In general, the higher the vapor pressure, the more volatile a chemical compound and therefore the more likely the chemical is to exist in significant quantities in a gaseous state. Thus, constituents with high vapor pressure are more likely to migrate from soil and groundwater, to be transported into air.

- *Boiling point.* Boiling point (BP) is the temperature at which the vapor pressure of a liquid is equal to the atmospheric pressure on the liquid. At this temperature, a substance transforms from the liquid into a vapor phase.

Besides being an indicator of the physical state of a chemical, the BP also provides an indication of its volatility. Other physical properties, such as critical temperature and latent heat (or enthalpy) of vaporization, may be predicted by use of a chemical's normal BP as an input.

3.2.6   Partitioning and Partition Coefficients

The partitioning of a chemical between several phases within a variety of environmental matrices is considered a very important fate and behavior property for contaminant migration from contaminated sites. The partition coefficient is a measure of the distribution of a given compound in two phases and is expressed as a concentration ratio. Several important measures of the partitioning phenomena are enumerated below.

- *Water/air partition coefficient.* The water/air partition coefficient (K_w) relates the distribution of a chemical between water and air. It consists of an expression that is equivalent to the reciprocal of Henry's Law constant (H), i.e.

$$K_w = \frac{C_{water}}{C_{air}} = \frac{1}{H}$$

where C_{air} is the concentration of the chemical in air (expressed in units of $\mu g/L$) and C_{water} is the concentration of the chemical in water (in $\mu g/L$).

- *Octanol/water partition coefficient.* The octanol/water partition coefficient (K_{ow}) is defined as the ratio of a chemical's concentration in the octanol phase (organic) to its concentration in the aqueous phase of a two-phase octanol/water system, represented by:

$$K_{ow} = \frac{\text{concentration in octanol phase}}{\text{concentration in aqueous phase}}$$

This dimensionless parameter provides a measure of the extent of chemical partitioning between water and octanol at equilibrium. It has become a particularly important parameter in studies of the environmental fate of organic chemicals.

K_{ow} can be used to predict the magnitude of an organic constituent's tendency to partition between the aqueous and organic phases of a two-phase system, such as surface water and aquatic organisms. The higher the value of K_{ow}, the greater the tendency of an organic constituent to adsorb to soil or waste matrices containing appreciable organic carbon or to accumulate in biota. It has been found to be related to water solubility, soil/sediment adsorption coefficients, and bioaccumulation factors for aquatic life. High K_{ow} values are generally indicative of a chemical's ability to accumulate in fatty tissues and therefore bioaccumulate in the foodchain. It is also a key variable in the estimation of skin permeability for chemical constituents.

In general, chemicals with low K_{ow} (<10) values may be considered relatively hydrophilic, whereas those with high K_{ow} (>10 000) values are very hydrophobic. Thus, the greater the K_{ow}, the more likely a chemical is to partition to octanol than to remain in water. The hydrophilic chemicals tend to have high water solubilities, small soil or sediment adsorption coefficients, and small bioaccumulation factors for aquatic life.

- *Organic carbon adsorption coefficient.* The sorption characteristics of a chemical may be normalized to obtain a sorption constant based on organic carbon which is essentially independent of any soil. The organic carbon adsorption coefficient (K_{oc}) provides a measure of the extent of partitioning of a chemical constituent between soil or sediment organic carbon and water at equilibrium. Also called the organic carbon partition coefficient, K_{oc} is a measure of the tendency for organics to be adsorbed by soil and sediment, and is expressed by:

$$K_{oc}[\text{mL/g}] = \frac{\mu g \text{ chemical adsorbed per g weight of soil or sediment organic carbon}}{\mu g \text{ chemical dissolved per mL of water}}$$

The extent to which an organic constituent partitions between the solid and solution phases of a saturated or unsaturated soil, or between runoff water and sediment, is determined by the physical and chemical properties of both the constituent and the soil (or sediment). The K_{oc} is chemical-specific and largely independent of the soil or sediment properties. The tendency of a constituent to be adsorbed to soil is, however, dependent on its properties and also on the organic carbon content of the soil or sediment.

Values of K_{oc} typically range from 1 to 10^7; the higher the K_{oc}, the more likely a chemical is to bind to soil or sediment than to remain in water. That is, constituents with a high K_{oc} have a tendency to partition to the soil or sediment. In fact, this value is also a measure of the hydrophobicity of a chemical; the more highly sorbed, the more hydrophobic (or the less hydrophilic) a substance.

- *Soil–water partition coefficient.* The mobility of contaminants in soil depends not only on properties related to the physical structure of the soil, but also on the extent to which the soil material will retain, or adsorb, the pollutant constituents. The extent to which a constituent is adsorbed depends on the physico-chemical properties of the chemical constituent and of the soil. The sorptive capacity must therefore be determined with reference to a particular constituent and soil pair.

The soil–water partition coefficient (K_d), also called the soil/water distribution coefficient, is generally used to quantify soil sorption. K_d is the ratio of the adsorbed contaminant concentration to the dissolved concentration at equilibrium, and for most environmental concentrations, it can be approximated by:

$$K_d[\text{mL/g}] = \frac{\text{concentration of adsorbed chemical in soil } (\mu g \text{ chemical per g soil})}{\text{concentration of chemical in solution in water } (\mu g \text{ chemical per mL water})}$$

K_d provides a soil- or sediment-specific measure of the extent of chemical partitioning between soil or sediment and water, unadjusted for dependence on organic carbon.

On this basis, K_d describes the sorptive capacity of the soil and allows estimation of the concentration in one medium, given the concentration in the adjoining medium. For hydrophobic contaminants: $K_d = f_{oc} K_{oc}$, where f_{oc} is the fraction of organic carbon in the soil.

In general, the higher the value of K_d, the less mobile is a contaminant; this is because, for large values of K_d, most of the chemical remains stationary and attached to soil particles due to the high degree of sorption. Thus, the higher the K_d the more likely a chemical is to bind to soil or sediment than to remain in water. Invariably, the distribution of a chemical between water and adjoining soil or sediment may be described by this equilibrium expression that relates the amount of chemical sorbed to soil or sediment to the amount in water at equilibrium.

- *Bioconcentration factor.* The bioconcentration factor (BCF) is the ratio of the concentration of a chemical constituent in an organism or whole body (e.g. a fish) or specific tissue (e.g. fat) to the concentration in its surrounding medium (e.g. water) at equilibrium, given by:

$$\text{BCF} = \frac{(\text{concentration in biota})}{(\text{concentration in surrounding medium})}$$
$$= \frac{[(\mu g \text{ chemical per g biota } <\text{e.g. fish}>)]}{[(\mu g \text{ chemical per mL medium } <\text{e.g. water}>)]}$$

The BCF indicates the degree to which a chemical residue may accumulate in aquatic organisms, coincident with ambient concentrations of the chemical in water; it is a measure of the tendency of a chemical in water to accumulate in the tissue of an organism. In this regard, the concentration of the chemical in the edible portion of the organism's tissue can be estimated by multiplying the concentration of the chemical in surface water by the fish BCF for that chemical. Thus, the average concentration in fish or biota is given by:

$$C_{\text{fish-biota}}(\mu g/kg) = C_{\text{water}}(\mu g/L) \times \text{BCF}$$

where C_{water} is the concentration in water. This parameter is indeed an important determinant for human exposure to chemicals via ingestion of aquatic foods. The partitioning of a chemical between water and biota (e.g. fish) also gives a measure of the hydrophobicity of the chemical.

Values of BCF typically range from 1 to over 10^6. In general, constituents exhibiting a BCF greater than unity are potentially bioaccumulative, but those exhibiting a BCF greater than 100 cause the greatest concern (USEPA, 1987a). Ranges of BCFs for various constituents and organisms can be used to predict the potential for bioaccumulation, and therefore to determine whether sampling of the biota is a necessary part of a site characterization program. The accumulation of chemicals in aquatic organisms is indeed of increased concern as a significant source of environmental and health hazard.

3.2.7 Sorption and the Retardation Factor

Sorption, which collectively accounts for both adsorption and absorption, is the partitioning of a chemical constituent between the solution and solid phases. In this partitioning process, molecules of the dissolved constituents leave the liquid phase and attach to the solid phase; this partitioning continues until a state of equilibrium is reached. The practical result of the partitioning is a phenomenon called retardation, in which the effective velocity of the chemical constituents is less than that of a 'pure' groundwater flow.

Retardation is the chemical-specific, dynamic process of adsorption to, and desorption from, aquifer materials. It is typically characterized by a parameter called the retardation factor or retardation coefficient. In the assessment of the environmental fate and transport properties of chemical contaminants, reversible equilibrium and controlled sorption may be simulated by the use of the retardation factor or coefficient.

- *Retardation factor.* The retardation factor, R_f, can be calculated for a contaminant as a function of the chemical's soil–water partition coefficient (K_d), and also the bulk density (β) and porosity (n) of the medium through which the contaminant is moving. Typically, the retardation factors are calculated for linear sorption, in accordance with the following relationship:

$$R_f = 1 + \frac{\beta K_d}{n} = \left[1 + \frac{\beta K_{oc} f_{oc}}{n}\right]$$

where $K_d = K_{oc} \times f_{oc}$, and f_{oc} is the organic carbon fraction.

The velocity of a contaminant is one of the most important variables in any groundwater quality modeling study. Sorption affects the solute seepage velocity through retardation, which is a function of R_f. Estimating R_f is therefore very important if solute transport is to be adequately represented. In the aquifer system, the retardation factor gives a measure of how fast a compound moves relative to groundwater (Nyer, 1993). Defined in terms of groundwater and solute concentrations, therefore,

$$R_f = \frac{\text{groundwater velocity } [v]}{\text{solute velocity } [v^*]}$$

For example, a retardation factor of two indicates that the specific compound is traveling at one-half the groundwater flow rate. This will usually become a very important parameter in the design of groundwater remediation systems. In particular, sorption can have major effects on pump-and-treat cleanup times and volumes of water to be removed from a contaminated aquifer system.

- *Sorption.* Under equilibrium conditions, a sorbing solute will partition between the liquid and solid phases according to the value of R_f. The fraction of the total contaminant mass contained in an aquifer which is dissolved in the solution phase, $F_{\text{dissolved}}$, and the sorbed fraction, F_{sorbed}, can be calculated as follows:

$$F_{\text{dissolved}} = \frac{1}{R_f}$$

$$F_{\text{sorbed}} = 1 - \left[\frac{1}{R_f}\right]$$

In general, if a compound is strongly adsorbed, then it also means this particular compound will be highly retarded.

3.2.8 Degradation

Degradation, whether biological, physical or chemical, is often reported in the literature as a half-life, which is generally measured in days. It is usually expressed as the time it takes for one-half of a given quantity of a compound to be degraded. Several important measures of the degradation phenomena are enumerated below.

- *Chemical half-lives.* Half-lives are used as measures of persistence, since they indicate how long a chemical will remain in various environmental media; long half-lives (e.g. greater than a month or a year) are characteristic of persistent constituents. Media-specific half-lives provide a relative measure of the persistence of a chemical in a given medium, although actual values can vary greatly depending on site-specific conditions. For example, the absence of certain microorganisms at a site, or the number of microorganisms, can influence the rate of biodegradation, and therefore the half-life for specific compounds. As such, half-life values should be used only as a general indication of a chemical's persistence in the environment. In general, however, the higher the half-life value, the more persistent a chemical is likely to be.
- *Biodegradation.* Biodegradation is one of the most important environmental processes affecting the breakdown of organic compounds. It results from the enzyme-catalyzed transformation of organic constituents, primarily by microorganisms. As a result of biodegradation, the ultimate fate of a constituent introduced into several environmental systems (e.g. soil, water, etc.) may be any compound other than the parent compound that was originally released into the environment. Biodegradation potential should therefore be carefully evaluated in the design of environmental monitoring programs in particular, and site assessments in general. It is noteworthy that biological degradation may also initiate other chemical reactions, such as oxygen depletion in microbial degradation processes, creating anaerobic conditions and the initiation of redox-potential-related reactions.
- *Photolysis.* Photolysis (or photodegradation) can be an important dissipative mechanism for specific chemical constituents in the environment. Similar to biodegradation, photolysis may cause the ultimate fate of a constituent introduced into an environmental system (e.g. surface water, soil, etc.) to be different from the constituent originally released. Hence, photodegradation potential should be carefully evaluated in designing sampling and analysis, as well as environmental monitoring programs.
- *Chemical degradation.* Similar to photodegradation and biodegradation, chemical degradation, primarily through hydrolysis and oxidation/reduction (redox) reactions, can also act to change chemical constituent species from what the parent compound used to be when it was first introduced to the environment. For instance, oxidation may occur as a result of chemical oxidants being formed during photochemical processes in natural waters. Similarly, reduction of constituents may take place in some surface water environments (primarily those with low oxygen levels). Hydrolysis of organics usually results in the introduction of a hydroxyl group (-OH) into a constituent structure; hydrated metal ions (particularly those with a valence $\geqslant 3$) tend to form ions in aqueous solution, thereby enhancing species solubility.

3.3 CONTAMINANT FATE AND TRANSPORT ASSESSMENT

Environmental contamination can be transported far away from its primary source(s) of origination via natural erosional and gravitational processes, resulting in the possible birth of new contaminated site problems. On the other hand, some natural processes work to lessen or attenuate contaminant concentrations in the environment through mechanisms of natural attenuation (such as dispersion/dilution, sorption and retardation, photolysis, and biodegradation). Typically, environmental fate and transport analysis and modeling are used to assess the movement of chemicals between environmental compartments. For instance, simple mathematical models can be used to guide the decisions involved in estimating the potential spread of contaminant plumes; where applicable, wells or monitoring equipment can then be located in areas expected to have elevated contaminant concentrations and/or in areas considered upgradient and downgradient of a contaminant plume.

3.3.1 Factors Affecting Contaminant Migration

Chemicals released into the environment are affected by several complex processes and phenomena that facilitate cross-media transfers. The affinity that contaminants have for soils can particularly affect their mobility by retarding transport. For instance, hydrophobic or cationic contaminants that are migrating in solution are subject to retardation effects. In fact, the hydrophobicity of a contaminant can greatly affect its fate, which explains some of the different rates of contaminant migration occurring in the subsurface environment. Also, the phenomenon of adsorption is a major reason why the sediment zones of surface water systems may become highly contaminated with specific organic and inorganic chemicals. The processes and phenomena that affect the fate and transport of contaminants should therefore be recognized as an important part of any diagnostic assessment and characterization program for contaminated sites.

Contaminant characteristics The physical and chemical characteristics of constituents present at contaminated sites determine the fate and transport properties of the contaminants, and thus their degree of migration through the environment. Some of the particularly important constituent properties affecting the fate and transport of contaminants in the environment include the following:

- Solubility in water (which relates to leaching, partitioning, and mobility in the environment).
- Partitioning coefficients (relating to cross-media transfers, bioaccumulation potential, and sorption by organic matter).
- Vapor pressure and Henry's Law constant (relating to atmospheric mobility and the rate of vaporization or volatilization).
- Degradation/half-life (relating to the degradation of contaminants and the resulting transformation products).
- Retardation factor (which relates to the sorptivity and mobility of the constituent within the solid–fluid media).

In the groundwater system, the solutes in the porous media will move with the mean velocity of the solvent by advective mechanism. In addition, other mechanisms governing the spread of contaminants include hydraulic dispersion and molecular diffusion. Furthermore, the transport and concentration of the solute(s) are affected by reversible ion exchange with soil grains; the chemical degeneration with other constituents; fluid compression and expansion; and in the case of radioactive materials, by the radioactive decay. The above-indicated parameters were discussed earlier in Section 3.2. Further details and additional parameters of possible interest are discussed elsewhere in the literature (e.g. Evans, 1989; Lyman et al., 1990; Swann and Eschenroeder, 1983; USEPA, 1989; Yong et al., 1992).

Site characteristics   Several site characteristics may also influence the environmental fate of chemicals, including the following in particular:

- Amount of ambient moisture, humidity levels, temperatures, and wind speed.
- Geologic, hydrologic, pedologic, and watershed characteristics.
- Topographic features of the site and its vicinity.
- Vegetative cover of site and surrounding area.
- Land-use characteristics.

Other factors such as soil temperature, soil moisture content, initial contaminant concentration in the impacted media, and media pH may additionally affect the release of a chemical constituent from the environmental matrix in which it is found.

Migration potential   In general, the degree of chemical migration from a contaminated site depends on both the physical and chemical characteristics of the individual constituents at the site, and also on the physical, chemical, and biological characteristics of the site. For example, physical characteristics of the contaminants such as solubility and volatility influence the rate at which chemicals leach into groundwater or escape into the atmosphere. The characteristics of the site environment (such as the geologic or hydrogeologic features) also affect the rate of contaminant migration. In addition, under various environmental conditions, some chemicals will readily degrade to substances of relatively low toxicity, while other chemicals may undergo complex reactions to become more toxic than the parent chemical constituent.

All other factors being equal however, the extent and rate of contaminant movement are a function of the physical containment of the chemical constituents or the contaminated zone. A classical illustration pertains to the fact that a low permeability cap over a contaminated site will minimize water percolation from the surface and therefore minimize leaching of chemicals into an underlying aquifer.

3.3.2 NAPL Fate and Transport

The process of determining the transport and fate of NAPLs can present very tough challenges, not only because of the many types of NAPLs and different kinds of soil materials that provide the medium through which the NAPLs interact, but also because of the mechanisms governing the fate of the NAPLs. The most important fluid properties that affect NAPL transport typically include volatility, relative polarity, affinity for soil organic matter or organic contaminants, density, viscosity, and interfacial tension (NRC, 1994; Yong et al., 1992). At the interface between groundwater and a NAPL pool, spreading of the dissolved components occurs primarily as a result of molecular diffusion.

Usually, LNAPLs are found near the top of the saturated zone, with the rate of transport depending on the local groundwater gradients and viscosity of the specific substance. DNAPLs, on the other hand, move downward through both the unsaturated and saturated zones, until they come to 'rest' at an impermeable boundary. DNAPLs generally have great penetration capability in the subsurface because of their relatively low solubility, high density and low viscosity (Yong et al., 1992). In fact, the relatively high density of these liquids provides a driving force that can carry contamination deep into aquifer systems that may exist at depth.

The mechanics of NAPL migration The most important mechanics of NAPL transport generally consist of aqueous transport within contaminant plumes, vapor-phase transport, and transport as a 'free product' (NRC, 1994). The relative importance of each of these for the transport of a particular contaminant depends on the properties of the substance involved as well as the site characteristics. For example, as soil moisture increases, the importance of vapor-phase transport diminishes; an increase in the moisture content reduces the pore space available for migration, decreasing the effective diffusivity and gas-phase permeability (NRC, 1994).

In general, as NAPL enters the subsurface environment, the liquid may dissolve into the water in the pores, volatilize into the air in the pores, or remain behind in the pore spaces as an entrapped residual liquid. In the saturated zone, some of the chemicals from the NAPL will dissolve in the flowing groundwater, forming a contaminant plume to be transported further downgradient of the contamination source; the remainder of the NAPL will either float atop the water table (i.e. for LNAPL) or continue its downward migration (i.e. for DNAPL) until a relatively impermeable barrier or stratum is encountered where the NAPL will pool on top of the obstruction.

3.3.3 Modeling Contaminant Fate and Transport

Chemical contaminants entering the environment tend to be partitioned or distributed across various environmental compartments. A good prediction of contaminant concentrations in the various environmental media is essential to adequately characterize contaminated sites, the results of which can also be used to support site restoration decisions. Mathematical algorithms can be used to predict the potential for contaminants to migrate from a contaminated site into potential receptor locations or compliance boundaries.

Mathematical models, which are mathematical equations used to simulate and predict real events and processes, often serve as valuable tools for evaluating the behavior and fate of chemical constituents in various environmental media. Transport and fate of contaminants can be predicted through the use of various methods, ranging from simple mass-balance and analytical procedures to multi-dimensional numerical solution of coupled differential equations. Several models of interest, together with model selection criteria and limitations, are discussed elsewhere in the literature of exposure modeling (e.g. Asante-Duah, 1993; CCME, 1994; CDHS, 1986; Ghadiri and Rose, 1992; USEPA, 1985, 1987b, 1988a, 1988b). In general, the appropriateness of a particular model depends on the characteristics of the particular problem. Thus, the screening of models should be tied to the project goals. Indeed, the wrong choice of models could result in the generation of false information, with consequential negative impacts on any decision made thereof. Ultimately, the choice of appropriate fate and transport models that will give reasonable indications of the contaminant behavior will help produce a realistic conceptual representation of the site; this is important to the characterization of any contaminated site problems, which in turn is a prerequisite to developing reliable site restoration strategies.

3.3.3.1 Model selection

The effective use of models in contaminant fate and behavior assessment depends greatly on the selection of models most suitable for this purpose. Numerous model classification systems with different complexities exist in practice – broadly categorized as analytical or numerical models, depending on the degree of mathematical sophistication involved in their formulation. Analytical models are models with simplifying underlying assumptions, often sufficient and appropriate for well-defined systems for which extensive data are available, and/or for which the limiting assumptions are valid. Whereas analytical models may suffice for some situations, numerical models (with more stringent underlying assumptions) may be required for more complex configurations and complicated systems.

The choice of which model to use for specific applications is subject to numerous factors. Thus, simply choosing a more complicated model over a simple one will not necessarily ensure a better solution in all situations. In fact, since a model is a mathematical representation of a complex system, some degree of mathematical simplification usually must be made about the system being modeled. Data limitations must be weighted appropriately, since it usually is not possible to obtain all of the input parameters due to the complexity (e.g. anisotropy and nonhomogeneity) of natural systems.

Ultimately, the type of model selected will be dependent on the overall goal of the assessment, the complexity of the site, the type of contaminants of concern, and the nature of the impacted and threatened media that are being considered in the investigation. General guidance for the effective selection of models in contaminant migration studies is provided in the literature elsewhere (e.g. CCME, 1994; CDHS, 1990; DOE, 1987; USEPA, 1985, 1987b, 1988a, 1988b; Walton, 1984; Zirschy and Harris, 1986). In several site assessment situations, a 'ballpark' or 'order-of-magnitude' (i.e. a rough approximation) estimate of the contaminant behavior and fate is usually all that is required for most analyses, in which case simple analytical models usually will suffice.

3.4 REFERENCES

Asante-Duah, D.K. (1993). *Hazardous Waste Risk Assessment*. CRC Press/Lewis Publishers, Inc., Boca Raton, Florida.

Brooks, S.M. et al. (1995). *Environmental Medicine*. Mosby, Mosby-Year Book, Inc., St Louis, Missouri.

CCME (Canadian Council of Ministers of the Environment) (1994). *Subsurface Assessment Handbook for Contaminated Sites.* Canadian Council of Ministers of the Environment (CCME), The National Contaminated Sites Remediation Program (NCSRP), Report No. CCME-EPC-NCSRP-48E (March 1994), Ottawa, Ontario, Canada.

CDHS (California Department of Health Services) (1986). *The California Site Mitigation Decision Tree Manual.* California Department of Health Services, Toxic Substances Control Division, Sacramento, California.

CDHS (California Department of Health Services) (1990). *Scientific and Technical Standards for Hazardous Waste Sites.* Prepared by the California Department of Health Services, Toxic Substances Control Program, Technical Services Branch Sacramento, California.

Charbeneau, R.J., P.B. Bedient, and R.C. Loehr (eds) (1992). *Groundwater Remediation. Water Quality Management Library*, Vol. 8. Technomic Publishing Co., Inc., Lancaster, Pennsylvania.

DOE (US Department of Energy) (1987). *The Remedial Action Priority System (RAPS): Mathematical Formulations.* US Department of Energy, Office of Environment, Safety & Health, Washington, DC.

Evans, L.J. (1989). Chemistry of metal retention by soils. *Environ. Sci. Technol.* 23(9), 1047–1056.

Ghadiri, H. and C.W. Rose (eds) (1992). *Modeling Chemical Transport in Soils: Natural and Applied Contaminants.* CRC Press/Lewis Publishers, Inc., Boca Raton, Florida.

Lyman, W.J., W.F. Reehl and D.H. Rosenblatt (1990). *Handbook of Chemical Property Estimation Methods: Environmental Behavior of Organic Compounds.* American Chemical Society, Washington, DC.

NRC (National Research Council) (1994). *Alternatives for Ground Water Cleanup.* Committee on Ground Water Cleanup Alternatives. National Academy Press, Washington, DC.

Nyer, E.K. (1993). *Practical Techniques for Groundwater and Soil Remediation.* Lewis Publishers, Boca Raton, Florida.

Samiullah, Y. (1990). *Prediction of the Environmental Fate of Chemicals.* Elsevier Applied Science (in association with BP), London, UK.

Swann, R.L. and A. Eschenroeder (eds) (1983). *Fate of Chemicals in the Environment.* ACS Symposium Series 225, American Chemical Society, Washington, DC.

USEPA (US Environmental Protection Agency) (1985). *Modeling Remedial Actions at Uncontrolled Hazardous Waste Sites.* EPA/540/2-85/001 (April, 1985). Office of Emergency and Remedial Response, Washington, DC.

USEPA (US Environmental Protection Agency) (1987a). *RCRA Facility Investigation (RFI) Guidance.* EPA/530/SW-87/001, Washington, DC.

USEPA (US Environmental Protection Agency) (1987b). *Selection Criteria for Mathematical Models Used in Exposure Assessments: Surface Water Models.* EPA-600/8-87/042. Office of Health and Environmental Assessment, Washington, DC.

USEPA (US Environmental Protection Agency (1988a). *Selection Criteria for Mathematical Models Used in Exposure Assessments: Ground-Water Models.* EPA-600/8-88/075. Office of Health and Environmental Assessment, Washington, DC.

USEPA (US Environmental Protection Agency) (1988b). *Superfund Exposure Assessment Manual, Report No. EPA/540/1-88/001*, OSWER Directive 9285.5-1, USEPA, Office of Remedial Response, Washington, DC.

USEPA (US Environmental Protection Agency) (1989). *Risk Assessment Guidance for Superfund.* Vol. I – *Human Health Evaluation Manual* (Part A). EPA/540/1-89/002. Office of Emergency and Remedial Response, Washington, DC.

Walton, W.C. (1984). *Practical Aspects of Ground Water Modeling.* National Water Well Association.

Yong, R.N., A.M.O. Mohamed, and B.P. Warkentin (1992). *Principles of Contaminant Transport in Soils. Developments in Geotechnical Engineering*, Vol. 73. Elsevier Scientific Publishers BV, Amsterdam, The Netherlands.

Zirschy, J.H. and D.J. Harris (1986). Geostatistical analysis of hazardous waste site data. *ASCE J. Environ. Engnr.* 112(4).

Chapter Four

Conceptualization of contaminated sites

Conceptual models are usually developed for contaminated sites in order to help identify and document likely contaminant source areas, migration and exposure pathways, potential receptors, and how these elements are inter-connected. Typically, the conceptual site model (CSM) establishes a hypothesis about possible contaminant sources, contaminant fate and transport, and possible pathways of exposure to the populations potentially at risk.

The development of a comprehensive CSM is generally recommended as a vital part of the diagnostic assessment and characterization of contaminated sites. This is particularly true because CSMs generally provide a systematic and structured framework for characterizing possible threats posed by contaminated sites; the conceptualization also aids in the organization and analysis of basic information relevant to the site restoration decisions about a contaminated site. As site characterization activities progress, the CSM may have to be revised and used to direct the next iteration of sampling activities necessary to complete the site characterization efforts. The updated or finalized CSM is used to develop realistic exposure scenarios for the project site.

4.1 ELEMENTS OF A CONCEPTUAL SITE MODEL

Typically a CSM is developed from available site sampling data, historical records, aerial photographs, and hydrogeologic information about the site. Once synthesized, this information may be presented in several different forms such as map views (showing sources, pathways, receptors, and the distribution of contamination), cross-sectional views (illustrating sectional components hydrogeology), and/or tabular forms (summarizing and comparing contaminant concentrations against background thresholds, and/or regulatory or risk-based standards).

A pictorial diagram of the conceptual representation of a contaminated site is shown in Figure 4.1. In general, a typical CSM will incorporate the following basic elements:

- Identification of site contaminants.
- Characterization of the source(s) of contamination.
- Delineation of potential migration pathways.
- Identification and characterization of all populations and resources potentially at risk.
- Determination of the nature of inter-connections between contaminant sources, contaminant migration pathways, and potential receptors.

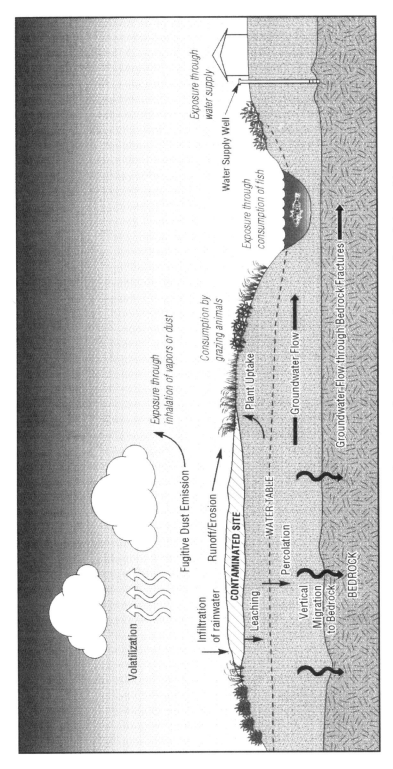

Figure 4.1   A diagrammatic conceptual representation of a contaminated site problem

Site history and preliminary assessment or site inspection data generally are very useful sources of information for developing preliminary CSMs. In general, the complexity and degree of sophistication of a CSM will be consistent with the complexity of the site and the amount of data available. In all cases, the CSM should include known and suspected sources of contamination, types of contaminants and affected media, known and potential migration pathways, potential exposure routes, target receiving media, and known or potential human and ecological receptors. Such information can be used to develop a conceptual understanding of the contaminated site problem, so that potential risks to human health and the environment can be evaluated more completely.

4.2   DESIGN OF CONCEPTUAL SITE MODELS

The development of an adequate CSM is an important aspect of a successful site characterization program. The CSM is also relevant to the development and evaluation of site restoration programs for contaminated sites. The design of a truly representative CSM for any contaminated site problem typically involves the following important evaluations:

- A contaminant release analysis (viz. determine the spatial distribution of contaminants; analyze site geology and hydrogeology; determine the extent to which contaminant sources can be adequately identified and characterized; determine the likelihood of releases if the contaminants remain on-site; determine the extent to which natural and artificial barriers currently contain contaminants, and the adequacy of such barriers; identify potential migration pathways; determine the extent to which site contaminants have migrated or are expected to migrate from their source(s); and estimate the contaminant release rates into specific environmental media over time).
- A contaminant transport and fate analysis (viz. provide guidance for evaluating the transport, transformation, and fate of contaminants in the environment following their release; identify off-site areas affected by contaminant migration; and determine contaminant amount, concentration, hazardous nature, and environmental fate properties of the constituents present in affected areas).
- An exposed population analysis (viz. determine the human, ecological and welfare resources potentially at risk; identify the routes of contaminant exposure to populations potentially at risk; determine the likelihood of human and ecological receptors coming into contact with the contaminants of concern; and assess the likelihood of contaminant migration posing a threat to public health, welfare, or the environment).
- An integrated exposure analysis (viz. provide guidance for calculating and integrating exposures to all populations affected by the various exposure scenarios associated with the contaminated site; and determine the extent to which contamination levels could exceed relevant regulatory standards, in relation to public health or environmental standards and criteria).

In general, the CSM should be appropriately modified if the acquisition of additional data and new information necessitates a re-design.

4.3 FROM CONCEPTUAL SITE MODELS TO THE DEVELOPMENT OF EXPOSURE SCENARIOS

The CSM facilitates an assessment of the nature and extent of contamination, and also helps determine the potential impacts from such contamination. Consequently, in as early a stage as possible during a site investigation, all available site information should be compiled and analyzed to develop a conceptual model for the site. This representation should incorporate contaminant sources and 'sinks', the nature and behavior of the site contaminants, migration pathways, the affected environmental matrices, and potential receptors. On this basis, a realistic set of exposure scenarios can be developed for a site characterization program.

An exposure scenario is a description of the activity that brings a population into contact with a contaminated environmental medium. Exposure scenarios are developed, based on the movement of contaminants in various environmental compartments (Figure 4.2). Several tasks are usually undertaken to facilitate the development of complete and realistic exposure scenarios; the critical tasks include the following:

• Determine the sources of site contamination.
• Identify the specific constituents of concern.
• Identify the affected environmental media.
• Delineate contaminant migration pathways.
• Identify potential receptors.
• Determine potential exposure routes.
• Construct a representative conceptual model for the site.
• Delineate likely and significant migration and exposure pathways.

The exposure scenario associated with a given contaminated site problem may be well defined if the exposure is known to have already occurred. In most cases associated with the investigation of potentially contaminated sites, however, decisions typically have to be made about potential exposures that may not yet have occurred. Consequently, hypothetical exposure scenarios are generally developed for such applications. Ultimately, the exposure scenarios developed for a given contaminated site problem can be used to support an evaluation of the risks posed by the site, as well as facilitate the implementation of appropriate decisions on the need for, and extent of, site restoration.

4.3.1 The Nature of Exposure Scenarios

A wide variety of *potential* exposure patterns can be anticipated from contaminated sites, indicating a multiplicity of inter-connected pathways through which populations

Primary and Secondary Sources → Migration and Exposure Pathways → Receptor Exposures

Figure 4.2 Exposure scenario evaluation flowchart

may be exposed to site contamination. The following represent typical and commonly encountered exposure scenarios in relation to contaminated site problems (Asante-Duah, 1993; HRI, 1995):

- Inhalation exposures
  - Indoor air, resulting from potential receptor exposure to contaminants (both volatile constituents and fugitive dust) in indoor ambient air.
  - Indoor air, resulting from potential receptor exposure to volatile chemicals in domestic water that may volatilize inside a house (e.g. during hot water showering) and contaminate indoor air.
  - Outdoor air, resulting from potential receptor exposure to contaminants (both volatile constituents and fugitive dust) in outdoor ambient air.
  - Outdoor air, resulting from potential receptor exposure to volatile chemicals in irrigation water, or other surface water bodies, that may volatilize and contaminate outdoor air.
- Ingestion exposures
  - Drinking water, resulting from potential receptor oral exposure to contaminants in domestic water used for drinking or cooking.
  - Swimming, resulting from potential receptor exposure (via incidental ingestion) to contaminants in surface water bodies.
  - Incidental soil ingestion, resulting from potential receptor exposure to contaminants in dust and soils.
  - Crop consumption, resulting from potential receptor exposure to contaminated foods (such as vegetables and fruits produced in household gardens that used contaminated soils, groundwater, or irrigation water in the cultivation process).
  - Dairy and meat consumption, resulting from potential receptor exposure to contaminated foods (such as locally grown livestock that may be contaminated from domestic water supplies, or from feeding on contaminated crops, or from contaminated air and soils).
  - Seafood consumption, resulting from potential receptor exposure to contaminated foods (such as fish and shellfish harvested from contaminated waters, or that have been exposed to contaminated sediments, and consequently have bioaccumulated toxic levels of chemicals in the edible portions).
- Dermal exposures
  - Showering, resulting from potential receptor exposure (via skin absorption) to contaminants in domestic water supply.
  - Swimming, resulting from potential receptor exposure (via skin absorption) to contaminants in surface water bodies.
  - Direct soils contact, resulting from potential receptor exposure to contaminants in outdoor soils.

These types of exposure scenarios will typically be evaluated as part of the site characterization process for a contaminated site problem. This listing is by no means complete, since new exposure scenarios are always possible for case-specific situations. In any case, once the complete set of potential exposure scenarios has been fully determined, the range of critical exposure pathways can be identified. This information can then be used to design cost-effective sampling and site investigation programs. The goal in this case will be to focus the site investigation, and to determine the specific potential exposure pathways associated with the planned use of the site. Subsequently, a range of engineering and/or institutional measures can be evaluated either to remove, to immobilize, or to otherwise control the site contaminants; or to mitigate the risks posed to potential human and ecological receptors associated with the site.

4.4 REFERENCES

Asante-Duah, D.K. (1993). *Hazardous Waste Risk Assessment.* CRC Press/Lewis Publishers, Inc., Boca Raton, Florida.
HRI (Hampshire Research Institute) (1995). *Risk*Assistant for Windows.* The Hampshire Research Institute, Inc., Alexandria, Virginia.

Chapter Five

Elements of a site characterization activity

The characterization of potentially contaminated sites is a process used to establish the presence or absence of contamination at a site, to delineate the nature and extent of site contamination, and to determine possible threats posed by the site to human health and/or the environment. Credible site characterization programs generally involve a complexity of activities that require careful planning. The initial step involves a data collection activity that is used to compile an accurate site description, history, and chronology of significant events. Subsequently, field sampling and laboratory analyses may be conducted to help define the nature and extent of contamination present at, or migrating from, a contaminated site.

A wide variety of investigation techniques may be employed in the characterization of contaminated sites. However, the appropriate methods and applicable techniques will generally be dependent on the type of contaminants, the site geologic and hydrogeologic characteristics, site accessibility, availability of several technical resources, and budget constraints. Several methods of choice used in conducting effective site characterization programs are described elsewhere in the literature of site assessment (e.g. BSI, 1988; CCME, 1993, 1994; CDHS, 1990; Driscoll, 1986; Lesage and Jackson, 1992; OBG, 1988; USEPA, 1985, 1987, 1988a, 1988b, 1989a, 1989b, 1989c; WPCF, 1988). Overall, the information obtained from a site characterization activity should be adequate to predict the fate and behavior of the contaminants in the environment, as well as to determine the potential impacts associated with the site and also to facilitate the design of an effectual corrective action program.

5.1 SCOPE OF A SITE CHARACTERIZATION ACTIVITY

The site characterization process consists of the collection and analysis of a variety of environmental data that is necessary for the design of an effectual site restoration program. The scope and detail of a site characterization activity should generally be adequate to determine the following:

- Primary and secondary sources of contamination
- Nature, amount and extent of contamination
- Fate and transport characteristics of site contaminants
- Pathways of contaminant migration
- Types of exposure scenarios associated with the site
- Risk to both human and ecological receptors/populations
- Feasible solutions to mitigate receptor exposures to site contaminants.

The completion of an adequate site characterization is indeed considered a very important component of any site restoration program that is designed to remedy effectively a contaminated site problem. The characterization of contaminant sources, migration pathways, and populations potentially at risk probably form the most important basis for determining the need for site remediation.

5.2 DEVELOPMENT OF FIELD INVESTIGATION WORKPLANS

As part of the site characterization program for a contaminated site, a carefully executed workplan should be developed to guide all relevant decisions. A typical workplan used to facilitate the investigation of contaminated site problems will generally consist of the following major components:

● A sampling and analysis plan
● A health and safety plan
● A waste management plan
● A quality assurance/quality control plan.

The important components and tasks required of most site investigation workplans are elaborated further in the following sections.

5.2.1 The Sampling and Analysis Plan

The sampling and analysis plan (SAP) is an essential requirement of any environmental site investigation. SAPs generally are required to specify sample types, numbers, locations and relevant procedures. Effective protocols are required in the sampling and laboratory procedures, in order to minimize uncertainties in the site investigation process. In a number of situations, the laboratory designated to perform the sample analyses provides sample bottles, preservation materials, and explicit sample collection instructions.

The principal objective of a sampling and analysis program is to obtain a small and informative portion of the statistical population being investigated so that contaminant levels can be established as part of a site characterization program. The methods by which data of adequate quality and quantity are to be obtained to meet the overall project goals should be specified and fully documented in the SAP that is developed as part of a detailed site characterization workplan. It is noteworthy that the selection of analytical methods is an integral part of the processes involved in the development of sampling plans, since this can strongly affect the acceptability of a sampling protocol. Furthermore, the use of appropriate sample collection methods can be as important as the use of appropriate analytical methods for sample analyses. Consequently, effective analytical protocols in both the sampling and laboratory procedures should be specified by the SAP, in order to help minimize uncertainties associated with the data collection and evaluation activities.

5.2.1.1 Elements of a sampling plan

Sampling protocols are written descriptions of the detailed procedures to be followed in collecting, packaging, labeling, preserving, transporting, storing, and documenting samples. Every sampling protocol must identify sampling locations, and should include all of the equipment and information needed for sampling. At a minimum, the following will be needed for documenting environmental sampling activities that are conducted at a potentially contaminated site (CCME, 1993; Keith, 1988, 1991): sampling date, sampling time, sample identification number, sampler's name, sampling site, sampling conditions or sample type, sampling equipment, preservation used, time of preservation, and relevant sample site observations (auxiliary data). In general, the devices used to collect, store, preserve, and transport samples must *not* alter the sample in any manner. Where necessary, special procedures should be employed to preserve samples during the period between collection and analysis. Several techniques and equipment that can be used in the sampling of contaminated soils, sediments, and waters are enumerated in the literature elsewhere (e.g. CCME, 1993; Keith, 1988, 1991).

Sampling considerations A history of the site, including the sources of contamination and a preliminary conceptual model describing the apparent migration and exposure pathways, should preferably be developed before a sampling plan is finalized. In certain situations, it may be necessary to conduct an exploratory study in order to give credence to the preliminary conceptual model. In general, as additional data identify specific areas where an initial conceptual site model is invalidated, the model should be modified to reflect the new information. An initial site evaluation usually provides insight into the types of site contaminants, the populations potentially at risk, and possibly the magnitude of the site risk. These particulars can be combined into the conceptual site model, which is subsequently used to guide the design of a sampling plan and to specify the size of sampling unit appropriate for the site characterization program.

Invariably all sampling plans should contain the following common elements:

- Site background (which includes a description of the site and surrounding areas and a discussion of known and suspected contamination sources, probable transport pathways, and other information about the site).
- Sampling objectives (describing the intended uses of the data).
- Sampling location and frequency (which also identifies each sample matrix to be collected and the constituents to be analyzed).
- Sample designation (which establishes a sample numbering system for the specific project, and should include the sample or well number, the sampling round, the sample matrix, and the name of the site).
- Sampling equipment and procedures (including equipment to be used and material composition of equipment, along with decontamination procedures).
- Sample handling and analysis (including identification of sample preservation methods, types of sampling jars, shipping requirements, and holding times).
- Quality assurance/quality control (QA/QC) requirements.

The key to a successful and cost-effective site characterization will be to optimize the number and length of field reconnaissance trips, the number of soil borings and wells, and the frequency of sampling and laboratory analyses. Opportunities to effect cost savings begin during the initial field investigation conducted as part of a preliminary assessment. In all situations, before a subsurface investigation is initiated, background information relating to the regional geology, hydrogeology, and site development history should have been fully researched. In general, minimizing of well installations and sample analyses during the initial phase of site characterization activities can greatly reduce overall costs. Indeed, well installation can even add to the spread of contamination by penetrating impermeable layers containing contaminants, and then allowing such contamination to migrate into previously uncontaminated deeper zones (Jolley and Wang, 1993). Consequently, non-intrusive and screening measurements are generally preferred modes of operations, whenever possible, since these tend to minimize both sampling costs and the potential to spread contamination. Ultimately, however, sufficient information will have to be collected to reliably show the identity, areal and vertical extents, and the magnitude of contamination associated with the project site.

Several important issues come into play when one is making a decision on how to obtain reliable samples: these include considerations of the sampling objective and approach; sample collection methods; chain-of-custody documentation; sample preservation techniques; sample shipment methods; sample holding times; and analytical protocols. A detailed discussion of sampling considerations and strategies for various environmental matrices can be found elsewhere in the literature (e.g. CDHS, 1990; Keith, 1988, 1991; USEPA, 1988b, 1989b).

5.2.1.2 Laboratory analytical protocols

The determination of analytical requirements involves specifying the most cost-effective analytical method which, together with the sampling methods, will meet the overall data quantity and quality objectives of a site investigation activity. Oftentimes, the initial analyses of environmental samples may be performed with a variety of field methods used for screening purposes. The rationale for using initial field screening methods is to decide if the level of pollution at a site is high enough to warrant more expensive (and more specific and accurate) laboratory analyses. Methods that screen for a wide range of compounds, even if determined as groups or homologues, are useful because they allow more samples to be measured faster and more inexpensively than with conventional laboratory analyses.

At a minimum, the following should be recorded as part of the requirements for documenting the laboratory work activities (CCME, 1993; USEPA, 1989c): method of analysis, date of analysis, laboratory or facility carrying out analysis, analyst's name, calibration charts and other measurement charts, method detection limits, confidence limits, records of calculations, and actual analytical results. Ultimately, in general, effective analytical programs and laboratory procedures are necessary to help minimize uncertainties in site investigation activities that are required to support contaminated site characterization programs as well as site restoration decisions. Guidelines for the selection of appropriate analytical methods are offered elsewhere in the literature (e.g. CCME, 1993; Keith, 1991; USEPA, 1989c). Invariably, analytical protocol and constituent parameter selection are usually carried out in a way that balances costs of analysis with adequacy of coverage.

5.2.2   The Health and Safety Plan

Contaminated sites, by their nature and definition, contain concentrations of chemicals that may be harmful to various human population groups. One significant group potentially at risk from site contamination is the field crew who enter the contaminated site to collect samples and/or to monitor the extent of contamination. To minimize risks to site workers that could result from exposure to site contamination, health and safety issues must always be addressed as part of the field investigation activity plan. Protecting the health and safety of the field investigation team, as well as the general public in the vicinity of the site, is indeed a major concern during the investigation of contaminated sites. Proper planning and execution of safety protocols will help protect the site investigation team from accidents and needless exposure to hazardous or potentially hazardous chemicals.

Purpose and scope of a health and safety plan   The purpose of a health and safety plan (HSP) is to identify, evaluate, and control health and safety hazards, and to provide for emergency response during fieldwork activities at a contaminated site. The HSP specifies safety precautions needed to protect the populations potentially at risk during on-site activities. Consequently, a site-specific HSP should be prepared and implemented prior to the commencement of any fieldwork activity at potentially contaminated sites. All personnel entering the project site will generally have to comply with the applicable HSP. Also, the scope and coverage of the HSP may be modified or revised to encompass any changes that may occur at the site, or in the working conditions, following the development of the initial HSP.

Overall, the HSP should be developed to conform with all the requirements for occupational safety and health, and also with applicable national, state/provincial/ regional and local laws, rules, regulations, statutes, and orders, as necessary to protect all populations potentially at risk. Furthermore, all personnel involved with on-site activities would have received adequate training, and there should be a contingency plan in place that meets all safety requirements.

Emergency phone numbers should be compiled and included in the HSP. Also, the directions to the nearest hospital or medical facility, including a map clearly showing the shortest route from the site to the hospital or medical facility, should be kept with the HSP at the site.

5.2.2.1   Implementation of the health and safety plan

Several relevant elements of a site-specific HSP that will satisfy the general requirements of a safe work activity are elaborated elsewhere in the literature (e.g. CDHS, 1990; USEPA, 1987). In general, every project should start with a health and safety review (at which all site personnel sign a review form), a tailgate safety meeting (to be attended by all site activities personnel), and a safety compliance agreement (which should be signed by all persons entering the site, i.e. both the fieldwork crew and site visitors). Contractors, subcontractors and other site investigation groups or teams, under the auspices of a 'Health and Safety Officer' (HSO), are required to implement the HSP during fieldwork activities.

Job functions of the HSO   The HSO has the primary responsibility for ensuring that the policies and procedures of the HSP are fully implemented. In particular, the designated HSO should take steps to protect personnel engaged in site activities from such physical hazards as falling objects (e.g. tools or equipment); tripping over hoses, pipes, tools, or equipment; slipping on wet or uneven surfaces; insufficient or faulty protective equipment; insufficient or faulty tools and equipment; overhead or below-ground electrical hazards; heat stress and strain; insect and reptile bites; inhalation of dust, etc. The designated HSO should also monitor and check the work habits of the site crew to ensure that they are safety conscious.

The HSO is responsible for providing copies of the HSP to the site crew (including subcontractors) and for advising the site crew on all health and safety matters. The HSO is also responsible for providing the appropriate safety monitoring equipment, and other resources necessary to implement the HSP. Significant deviations/changes to the original HSP must be approved by the HSO. Typically, the HSO will have the authority to resolve outstanding health and safety issues that come up during the site investigation activities. The HSO must have stop-work authority if a dangerous or potentially dangerous and unsafe situation exists at the site. Consequently, the HSO must be notified of any changes in site conditions, during the course of a site characterization activity, which may have significant impacts on the safety of personnel, the environment, or property.

5.2.2.2   Health and safety requirements in the investigation of contaminated sites

Several health and safety requirements are generally adopted during the investigation of potentially contaminated site problems. Health and safety data are often required to establish the level of protection needed for the site investigation crew entering a potentially contaminated site. Such data are also used to determine if there should be immediate concern for any population living in proximity of the site. In general, safety plans should include requirements for hard hats, safety boots, safety glasses, respirators, self-contained breathing apparatus, gloves, and hazardous materials suits, if needed. The use of appropriate types of protection equipment will ensure the safety of the field crew. In addition, personal exposure monitoring and/or monitoring ambient air concentrations of some chemicals may be necessary to meet safety regulations. In all cases, the HSO establishes the level of protection required and determines whether the level should be advanced or reduced.

Personnel protection equipment   The level of personnel protection equipment (PPE) to be worn by field personnel is identified and enforced by the designated HSO. The levels of protection may change as additional information is acquired. For most of the typical site activities conducted at potentially contaminated sites, direct worker contact with hazardous materials in soil can be mitigated by using the lowest level of personal protective equipment – consisting of cover-alls, safety boots, glasses, and a hard hat. To protect workers from unacceptable levels of airborne materials, at least a level of protection that includes a full-face-piece air-purifying respirator may be required. In certain other cases, worker exposure to toxic materials may be such as to warrant even more sophisticated equipment, in order to provide yet greater levels of protection against exposure. In general, if a higher level of protection above that originally specified in the HSP is needed, then approval by the HSO will be required. It is noteworthy that all personnel utilizing most high level PPEs must have successfully passed a thorough physical examination, and should indeed have received the consent of a qualified physician prior to engaging in any on-site activities that utilize these types of equipment.

Decontamination   All site personnel as well as equipment used for site-related activities should be subject to a thorough decontamination process. Separate decontamination stations should preferably be established for personnel, equipment, and other machinery. The decontamination area should be clearly delineated, highly visible, and accessible to all personnel engaged in site activities.

Health and safety training   All personnel working at a project site should have participated in adequate health and safety training. This should be ascertained and certified by the designated HSO before an individual is allowed to enter into and work at a potentially contaminated site. The HSO should also conduct on-site health and safety training covering items required by the HSP. Additional safety briefings should be provided by the HSO if the scope of work changes in a manner that will potentially affect personal health and safety of the workers. All personnel engaged in site activities are required to participate in the training by the HSO. All persons involved in these activities should sign a health and safety review sheet, certifying that they have read, understand, and agree to comply with the stipulations of the HSP.

Emergency response   Several provisions for an emergency response plan should be carried out whenever there is a fire, explosion, or release of hazardous material which could threaten human health or the environment. If a situation requires outside assistance, the appropriate response parties should be contacted using mobile/cellular phones to be carried to the site, and/or phones located in nearby facilities that should have been identified at the start of the site activities.

In all of the emergency response situations, the emergency transport and medical personnel should immediately be notified as to the type and degree of injury, as well as the extent and nature of contamination to the injured or affected individual(s).

5.2.3   The Waste Management Plan

Investigation-generated wastes (IGWs) are those wastes produced during site characterization and remedial activities; examples include drill cuttings or core samples from soil boring or monitoring well installations, drilling muds, well purging water, and decontamination water. The objective of the waste management plan is to specify procedures needed to address and handle both hazardous and non-hazardous IGWs. Specifically, the IGW management plan should include the characterization of the IGW, delineation of any areas of contamination, and the identification of waste disposal methods. Ultimately, the site manager should select investigation methods that minimize the generation of IGWs. Minimizing the amount of wastes generated during a site characterization activity generally reduces the number of IGW handling problems and costs for disposal. Insofar as possible, provisions should be made for the proper handling and disposal of IGWs on-site. To handle IGWs properly, the site manager must, among other things, determine the waste types, the waste characteristics, and the quantities of anticipated wastes. To the extent practicable, the handling, storage, treatment, or disposal of any IGWs produced during site characterization and remedial activities must satisfy all regulatory requirements and stipulations that are applicable or relevant and appropriate to the specific project.

5.2.4 The Quality Assurance/Quality Control Plan

Quality assurance (QA) refers to a system for ensuring that all information, data, and resulting decisions compiled from an investigation (e.g. monitoring and sampling tasks) are technically sound, statistically valid, and properly documented. Quality control (QC) is the mechanism through which quality assurance achieves its goals. A detailed QA/QC plan, describing specific requirements for the QA and QC of both laboratory analysis and field sampling/analysis, should be part of the site characterization project workplan. Typically, some aspects of the field program are subjected to a quality assessment survey, accomplished by submitting sample blanks (alongside the environmental samples) to the laboratory for analysis on a regular basis. The various blanks and checks that are usually recommended as part of the QA/QC plan include the following (CCME, 1994):

- *Trip blank* – required to identify contamination of bottles and samples during travel and storage.
- *Field blank* – required to identify contamination of samples during collection.
- *Equipment blank* – required to identify contamination from well and sampling equipment.
- *Blind replicate* – required to identify laboratory variability.
- *Spiked sample* – required to identify errors due to sample storage and analysis.

Since data generated during a site characterization will provide a basis for site restoration decisions, such data should give a valid representation of the true site conditions. The development and implementation of a credible QA/QC program as part of a site characterization program is critical to obtaining reliable analytical results. The soundness of the QA/QC program has a particularly direct bearing on the integrity of the environmental sampling and also the laboratory work. Thus, the general design process for an adequate QA/QC program, as discussed elsewhere in the literature (e.g. CCME, 1994; USEPA, 1987, 1988a), should be adhered to in the strictest manner.

5.3 IMPLEMENTATION OF A SITE CHARACTERIZATION ACTIVITY PROGRAM

Site investigations are conducted in order to characterize conditions at a project site. Environmental samples collected from the site will generally be submitted to a certified analytical laboratory for analysis. Essentially, the characterization of a contaminated site helps determine the specific type(s) of site contaminants, their abundance or concentrations, the lateral and vertical extents of contamination, the volume of contaminated materials involved, and the background contaminant levels for native soils and water resources in the vicinity of the site.

As an important starting point in the site investigation process, the quality of data required from the specific study should be clearly defined. Once the level of confidence required for site data is established, strategies for sampling and analysis can then be developed (USEPA, 1988a). The identification of sampling requirements involves specifying the sampling design; the sampling method; sampling numbers, types, and locations; and the level of sampling quality control. In fact, sampling program designs must seriously consider the quality of data needed. If the samples are not collected, preserved, and stored correctly before they are analyzed, the analytical data may be compromised. Also, if sufficient sample amounts are not collected, the method sensitivity requirements may not be achieved.

Appendix C of this title summarizes several important elements and requirements that should be duly considered during the implementation stage of a site characterization activity program for any potentially contaminated site problem.

5.3.1   An Initial Site Inspection

Geophysical surveys, limited field screening, or limited field analyses may be performed during an initial site inspection. These types of preliminary screening activities may help determine the variability of the environmental media, provide general interest background information, or determine if the site conditions have changed in comparison to what may have been reported in previous investigations. Typically, the goals of the initial site inspection consist of completing the following tasks:

- Utilizing field analytical procedures, compile data on volatile chemical contaminants, radioactivity, and explosivity hazards in order to determine appropriate health and safety level requirements.
- Update site conditions if undocumented changes have occurred, and determine if any site condition could pose an imminent danger to public health and safety.
- Confirm, insofar as possible, all relevant information contained in previous documents and record any apparent discrepancies and/or any observable data that may be missing in previous documents.
- Investigate and inventory all possible off-site sources that may be contributing to contamination at the project site.
- Obtain information on location of access routes, sampling points and site organizational requirements for the field investigation.

Available information should be carefully reviewed and evaluated to provide the foundation for executing on-site activities. The type of information generated serves as a useful database for project scoping. The review and initial site visit are used in a preliminary interpretation of site conditions.

5.3.2   A Strategy for the Field Investigation Activities

A preliminary identification of the nature of site contaminants, the chemical release potentials, and also the likely migration and exposure pathways should be made very early in the site characterization, because these are crucial to decisions on the number, type, and location of samples to be collected. In fact, the nature of chemicals believed or known to be present at a contaminated site may dictate the areas and extent of environmental media to be sampled. For instance, in the design of a sampling program, if it is believed that the contaminants of concern are relatively immobile (e.g. most metals in silty or clayey material), then sampling may initially focus on soils in the vicinity of the suspected source of contaminant releases. On the other hand, the sampling design for a site believed to be contaminated by more mobile compounds (e.g. organic solvents in sandy materials) should necessarily take account of the fact that contamination may already have migrated into groundwater systems and/or moved significant distances away from the original source area(s). Consequently, knowledge on the type of contaminants will generally help focus more attention on the specific media most likely to be impacted. Similar decisions as above will typically have to be made regarding the choice of analytical protocols. For instance, due to the differences in the relative toxicities of the different species of some chemicals (e.g. chromium may exist as trivalent chromium, Cr^{3+}, or as the more toxic hexavalent chromium, Cr^{6+}), chemical speciation to differentiate between the various forms of the same chemical of potential concern present at a contaminated site may be required in the design of some analytical protocols.

Ideally, samples from various media should be collected in a manner that accounts for temporal factors and weather conditions. If seasonal fluctuations cannot be characterized in the investigation, details of meteorological, seasonal, and climatic conditions during sampling must be well documented. In fact, the ideal sampling strategy will incorporate a full annual sampling cycle. If this strategy cannot be accommodated in an investigation, at least two sampling events should be considered that take place during opposite seasonal extremes (such as high-flow/low-flow, high-recharge/low-recharge, etc.). It is noteworthy that, regardless of the medium sampled, data variability problems may arise from temporal and spatial variations in field data. That is, sample composition may vary depending on the time of the year and weather conditions when the sample is collected.

5.3.3 Sampling and Sample Handling

The collection of representative samples generally involves different procedures for different situations. Several sampling methods of approach that can be used in various types of situations are discussed elsewhere in the literature (e.g. CCME, 1994; CDHS, 1990; Keith, 1988, 1991; USEPA, 1985, 1987). In every situation, all sampling equipment (which should be constructed of inert materials) is cleaned using a non-phosphate detergent, a tap water rinse, and a final rinse with distilled water prior to a sampling activity. Decontamination of equipment is necessary so that sample results do not show false positives. Decontamination water generated from the site activities (e.g. during decontamination of hand auger and soil sampling equipment) usually is transferred into containers and treated as an IGW.

Soil samples collected are placed in resealable plastic bags; fluid samples are placed in air-tight glass or plastic containers. When samples are to be analyzed for organic constituents, glass containers are required. The samples are then labeled with an indelible marker. Each sample bag or container is labeled with a sample identification number, sample depth (where applicable), sample location, date and time of sample collection, preservation used (if any), and possibly a project number and the sampler's initials. A chain-of-custody form listing the sample number, date and time of sample collection, analyses requested, a project number, and persons responsible for handling the samples is then completed. Samples are generally kept on ice prior to and during transport/shipment to a certified or credible laboratory for analysis; completed chain-of-custody records should accompany all samples going to the laboratory.

The appropriate technical standards for sampling and sample handling procedures can be found in the literature elsewhere (e.g. CCME, 1994; CDHS, 1990; Keith, 1988, 1991; USEPA, 1985). In any case, all sampling should be conducted in a manner that maintains sample integrity, and should meet the requisite QA/QC criteria. The use of field blanks and standards, and also spiked samples, can account for changes in samples which occur after sample collection. Specific sample locations should be chosen such that representative samples can be collected. Also, samples should be collected from locations with visual observations of surface contamination, so that possible worst-case conditions may be identified.

5.4 REFERENCES

BSI (British Standards Institution) (1988). Draft for Development, DD175: 1988 Code of Practice for the Identification of Potentially Contaminated Land and its Investigation. BSI, London, UK.

CCME (Canadian Council of Ministers of the Environment) (1993). *Guidance Manual on Sampling, Analysis, and Data Management for Contaminated Sites*. Vol. I: *Main Report* (Report CCME EPC-NCS62E), and Vol. II: *Analytical Method Summaries* (Report CCME EPC-NCS66E) (December 1993). The National Contaminated Sites Remediation Program, Winnipeg, Manitoba.

CCME (Canadian Council of Ministers of the Environment) (1994). *Subsurface Assessment Handbook for Contaminated Sites*. Canadian Council of Ministers of the Environment (CCME), The National Contaminated Sites Remediation Program (NCSRP), Report No. CCME-EPC-NCSRP-48E (March 1994), Ottawa, Ontario, Canada.

CDHS (California Department of Health Services) (1990). *Scientific and Technical Standards for Hazardous Waste Sites*. Prepared by the California Department of Health Services, Toxic Substances Control Program, Technical Services Branch, Sacramento, California.

Driscoll, F.G. (1986). *Groundwater and Wells*. Johnson Division, St Paul, Minnesota.

Jolley, R.L. and R.G.M. Wang (eds) (1993). *Effective and Safe Waste Management: Interfacing Sciences and Engineering with Monitoring and Risk Analysis*. Lewis Publishers, Boca Raton, Florida.

Keith, L.H. (ed.) (1988). *Principles of Environmental Sampling*. American Chemical Society (ACS), Washington, DC.

Keith, L.H. (1991). *Environmental Sampling and Analysis – A Practical Guide*. Lewis Publishers, Boca Raton, Florida.

Lesage, S. and R.E. Jackson (eds) (1992). *Groundwater Contamination and Analysis at Hazardous Waste Sites*. Marcel Dekker, Inc., New York.

OBG (O'Brien & Gere Engineers, Inc.) (1988). *Hazardous Waste Site Remediation: The Engineer's Perspective*. Van Nostrand Reinhold, New York.

USEPA (US Environmental Protection Agency) (1985). *Characterization of Hazardous Waste Sites: A Methods Manual*. Vol. 1 – *Site Investigations*. US Environmental Protection Agency, Environmental Monitoring Systems Laboratory, Las Vegas, Nevada. EPA-600/4-84-075.

USEPA (US Environmental Protection Agency) (1987). *RCRA Facility Investigation (RFI) Guidance*. EPA/530/SW-87/001. Washington, DC.

USEPA (US Environmental Protection Agency) (1988a). *Guidance for Conducting Remedial Investigations and Feasibility Studies Under CERCLA.* EPA/540/G-89/004. OSWER Directive 9355.3-01, Office of Emergency and Remedial Response, Washington, DC.

USEPA (US Environmental Protection Agency) (1988b). *Interim Report on Sampling Design Methodology.* Environmental Monitoring Support Laboratory, Las Vegas, Nevada. EPA/600/X-88/408.

USEPA (US Environmental Protection Agency) (1989a). *Ground-water Sampling for Metals Analyses.* Office of Solid Waste and Emergency Response. EPA/540/4-89-001.

USEPA (US Environmental Protection Agency) (1989b). *Soil Sampling Quality Assurance User's Guide*, 2nd edn. EPA/600/8-89/046, Experimental Monitoring Support Laboratory (EMSL), ORD, US EPA, Las Vegas, Nevada.

USEPA (US Environmental Protection Agency) (1989c). *User's Guide to the Contract Laboratory Program.* Office of Emergency and Remedial Response, OSWER Dir. 9240.0-1.

WPCF (Water Pollution Control Federation) (1988). Hazardous Waste Site Remediation: Assessment and Characterization. A Special Publication of the WPCF, Technical Practice Committee, Alexandria, Virginia.

Chapter Six

Risk assessment as a diagnostic tool

Risk assessment is generally considered an integral part of the diagnostic assessment of potentially contaminated site problems. In most applications, it is used to provide a baseline estimate of existing risks that are attributable to a specific agent or hazard; the baseline risk assessment consists of an evaluation of the potential threats to human health and the environment in the absence of any remedial action. Risk assessment can also be used to determine the potential reduction in exposure and risk under various corrective action scenarios.

In its application to the investigation of potentially contaminated site problems, the risk assessment process encompasses an evaluation of all the significant risk factors associated with all feasible and identifiable exposure scenarios that are the result of contaminant releases into the environment. It involves the characterization of potential adverse consequences or impacts to human and ecological receptors that are potentially at risk from exposure to site contaminants.

Invariably, site restoration decisions about contaminated sites are made primarily on the basis of potential human health and ecological risks. A risk assessment process is generally utilized to determine whether the level of risk at a contaminated site warrants remediation, and to further project the amount of risk reduction necessary to protect public health and the environment. In particular, baseline risk assessments are usually conducted to evaluate the need for, and the extent of, remediation required at potentially contaminated sites. That is, they provide the basis and rationale as to whether or not remedial action is necessary. Overall, the baseline risk assessment contributes to the adequate characterization of contaminated site problems. It further facilitates the development, evaluation, and selection of appropriate corrective action response alternatives.

6.1   THE PURPOSE OF RISK ASSESSMENT

Risk assessment is a process used to determine and characterize possible risks that a potentially contaminated site poses to public health and the environment. Its general purpose is to gather sufficient information that will allow for an adequate and accurate characterization of potential risks associated with a project site. The risk assessment process integrates the information obtained during a site characterization or remedial investigation activity into a coherent set of goals for a feasibility study or the restoration of contaminated sites. Several advantages may indeed accrue from the use of risk assessment procedures to arrive at consistent and cost-effective corrective action decisions for contaminated site problems.

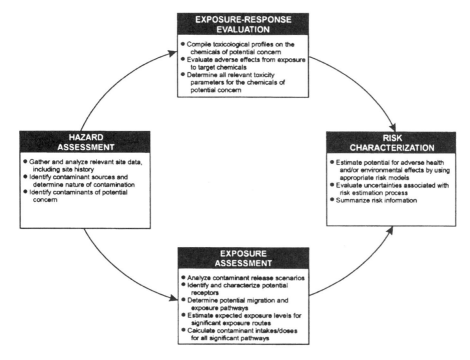

Figure 6.1 Fundamental elements of the risk assessment process

Oftentimes, the information generated in a risk assessment is used to determine the need for, and the degree of, remediation required for potentially contaminated sites. Risk assessment techniques and principles can generally be employed to facilitate the development of effectual site characterization and corrective action programs. In addition to providing information about the nature and magnitude of potential health and environmental risks associated with a contaminated site problem, risk assessment also provides a basis for judging the need for remediation. Furthermore, risk assessment can be used to compare the risk reductions afforded by different remedial or risk control strategies.

6.2 THE RISK ASSESSMENT PROCESS

Specific forms of risk assessment generally differ considerably in their levels of detail. Most risk assessments, however, share the same general logic, consisting of four basic elements (Figure 6.1). A discussion of these fundamental elements follows, with more detailed elaboration given elsewhere in the risk analysis literature (e.g. Asante-Duah, 1990, 1993; Bowles et al., 1987; CAPCOA, 1990; Hallenbeck and Cunningham, 1988; Huckle, 1991; NRC, 1983; Paustenbach, 1988; Rowe, 1977; USEPA, 1984, 1989a, 1989b).

In the investigation of contaminated site problems, the focus of most risk assessments is on a determination of potential risks to human and ecological receptors. Although these represent different types of populations, the mechanics of the evaluation process are similar. It is noteworthy that, in general, much of the effort in the development of risk assessment methodologies has been directed at human health risk assessments (as reflected by the differences in the depth of coverage for human health versus ecological risk assessment that can be found in the literature). However, the fundamental components of the risk assessment process for other biological organisms parallel those for human receptors, and can indeed be described in similar terms.

Hazard assessment   Hazard assessment involves a qualitative assessment of the presence of, and the degree of, hazard that a contaminant could have on potential receptors. In the context of the diagnosis of contaminated site problems, this may consist of the identification of contaminant sources; a compilation of the lists of all contaminants present at the site; the identification and selection of the specific chemicals of potential concern (which should become the focus of the risk assessment), based on their specific hazardous properties (such as persistence, bioaccumulative properties, toxicity, and general fate and transport properties); and a compilation of summary statistics for the key constituents selected for further investigation and evaluation.

Exposure–response evaluation   The exposure–response evaluation will generally include a toxicity assessment and/or a dose–response evaluation. It considers the types of adverse effects associated with chemical exposures, the relationship between magnitude of exposure and adverse effects, and related uncertainties (such as the weight-of-evidence of a particular chemical's carcinogenicity in humans). The toxicity assessment consists of compiling toxicological profiles for the chemicals of potential concern. Dose–response relationships are then used to evaluate quantitatively the toxicity information and to characterize the relationship between dose of the contaminant administered or received and the incidence of adverse effects on the exposed population. From the quantitative dose–response relationship, appropriate toxicity values can be derived and subsequently used to estimate the incidence of adverse effects occurring in populations at risk for different exposure levels.

Exposure assessment   An exposure assessment is conducted to estimate the magnitude of actual and/or potential receptor exposures to environmental contaminants, the frequency and duration of these exposures, the nature and size of the populations potentially at risk, the contaminant migration pathways, and the routes by which the risk group may be exposed. To complete a typical exposure assessment, populations potentially at risk are identified, and concentrations of the chemicals of concern are determined in each medium to which potential receptors may be exposed. Finally, using the appropriate site-specific exposure parameter values, the intakes of the chemicals of potential concern are estimated. The exposure estimates can then be used to determine if any threats exist based on existing exposure conditions at or near the contaminated site.

Risk characterization   Risk characterization is the process of estimating the probable incidence of adverse impacts to potential receptors under a set of exposure conditions. Typically, the risk characterization summarizes and then integrates outputs of the exposure and toxicity assessments in order to qualitatively and/or quantitatively define risk levels. This usually will include an elaboration of uncertainties associated with the risk estimates. Exposures resulting in the greatest risk can be identified in this process; site mitigation measures can then be selected to address the situation in order of priority, and according to the levels of imminent risks. In fact, an adequate characterization of risks from hazards at a contaminated site allows site restoration decisions to be better focused.

6.2.1 Human Health Risk Assessment

Human health risk assessment is defined as the characterization of the potential adverse health effects associated with human exposures to environmental hazards (NRC, 1983). Quantitative human health risk assessment is often an integral part of most site characterization as well as site restoration programs that are designed for contaminated site problems. The basic components and tasks involved in a comprehensive human health risk assessment consist of the following:

- Data evaluation
  - Assess the quality of available data.
  - Identify, quantify, and categorize site contaminants.
  - Screen and select chemicals of potential concern.
  - Carry out statistical analysis of relevant site data.
- Exposure assessment
  - Compile information on the physical setting of the site.
  - Identify source areas, significant migration pathways, and potentially impacted or receiving media.
  - Determine the important environmental fate and transport processes for the chemicals of potential concern, including cross-media transfers.
  - Identify populations potentially at risk.
  - Determine likely and significant receptor exposure pathways.
  - Develop representative conceptual site model(s).
  - Develop exposure scenarios (to include both the current and potential future land-uses for the site).
  - Estimate/model exposure point concentrations for the chemicals of potential concern found in the significant environmental media.
  - Compute potential receptor intakes and resultant doses for the chemicals of potential concern (for all potential receptors and significant pathways of concern).
- Toxicity assessment
  - Compile toxicological profiles (to include the intrinsic toxicological properties of the chemicals of potential concern, such as their acute, subchronic, chronic, carcinogenic, and reproductive effects).
  - Determine appropriate toxicity indices (such as the acceptable daily intakes or reference doses, cancer slope or potency factors, etc.).
- Risk characterization
  - Estimate carcinogenic risks from carcinogens.
  - Estimate noncarcinogenic hazard quotients and indices for systemic toxicants.
  - Perform sensitivity analyses, evaluate uncertainties associated with the risk estimates, and summarize the risk information.

Several important aspects of the human health risk assessment methodology are enumerated below; further details can be found elsewhere in the literature (e.g. Asante-Duah, 1993; Huckle, 1991; NRC, 1983; USEPA, 1989b, 1990a).

6.2.1.1 Data evaluation

The data evaluation aspect of a human health risk assessment consists of an identification and analysis of the chemicals present at a potentially contaminated site that should become the focus of the site characterization. In this process, an attempt is generally made to select all chemicals that could represent the major part of the risks associated with site-related exposures; typically, this will consist of all constituents contributing $\geqslant 95\%$ of the site risks. Chemicals are screened based on such parameters as toxicity, carcinogenicity, concentrations of the detected chemicals, and the frequency of detection in the sampled matrix.

6.2.1.2 Exposure assessment

The exposure assessment phase of the human health risk assessment is used to estimate the rates at which chemicals are absorbed by potential receptors. Since most potential receptors tend to be exposed to chemicals from a variety of sources and/or in different environmental media, an evaluation of the relative contributions of each medium and/ or source to total chemical intake could be critical in a multi-pathway exposure scenario. In fact, the accuracy with which exposures are characterized could be a major determinant of the ultimate validity of a risk assessment.

Exposure pathways are one of the most important elements of the exposure assessment process, consisting of the routes that contaminants follow to reach potential receptors. In fact, failure to identify and address any significant exposure pathway may seriously detract from the usefulness of any risk assessment. For receptor exposure to occur, a complete pathway must be present, which is used to define the exposure scenarios that will realistically exist for the case site. An exposure pathway is considered complete only if all on the following elements are present:

- Contaminant source(s)
- Mechanism(s) of contaminant release into the environment
- Contaminant migration pathway(s) and exposure route(s)
- Receptor exposure in the affected media.

Exposure pathways are determined by integrating information from an initial site characterization with knowledge about potentially exposed populations and their likely behavior. The significance of migration pathways is evaluated on the basis of whether the contaminant migration could cause significant adverse human exposures and impacts. The exposure routes (which may consist of inhalation, ingestion, and/or dermal contacts) and duration of exposure (which may be short term [acute] or long term [chronic]) will significantly influence the level of impacts on the affected receptors.

Estimation of receptor exposures   Several techniques may be used to perform the exposure assessment, including the modeling of anticipated future exposures; environmental monitoring of current exposures; and biological monitoring to determine past exposures. In general, once exposure point concentrations in all media of concern have been determined, the intakes (defined as the amount of chemical coming into contact with the receptor's body or exchange boundaries [such as the skin, lungs, or gastrointestinal tract]) and/or doses (defined as the amount of chemical absorbed by the body into the bloodstream) to potentially exposed populations can be estimated. The absorbed dose usually will differ significantly from the externally-applied dose (called exposure or intake).

Intakes and doses are normally calculated in the same step of the exposure assessment, whereby the former multiplied by an absorption factor yields the latter value. Potential receptor exposures to environmental contaminants can be conservatively estimated according to the generic equation shown in Box 6.1. The various exposure parameters are usually based on information relating to the maximum exposure level that results from specified categories of human activity and/or exposures (CAPCOA, 1990; OSA, 1993; USEPA, 1987, 1988, 1989a, 1989c, 1991). The methods by which each specific type of exposure is estimated, including the relevant equations for specific major routes of exposure, are included in Appendix D of this volume and further documented in greater detail elsewhere in the literature (e.g. Asante-Duah, 1993; CAPCOA, 1990; DOE, 1987; USEPA, 1988, 1989a, 1989c).

In general, determination of the carcinogenic effects (and sometimes the chronic noncarcinogenic effects) from a contaminated site involve estimating a lifetime average daily dose (LADD). For noncarcinogenic effects, an average daily dose (ADD) is usually used. The ADD differs from the LADD in that the former is not averaged over a lifetime; rather, it is the average daily dose pertaining to the number of days of actual

---

Box 6.1   General equation for estimating potential receptor exposures to site contaminants

$$EXP = \frac{(C_{medium} \times CR \times CF \times FI \times ABS_f \times EF \times ED)}{(BW \times AT)}$$

where:
EXP = intake (i.e. the amount of chemical at the exchange boundary), adjusted for absorption (mg/kg-day)

C_{medium} = average or reasonably maximum exposure concentration of chemical contacted by potential receptor over the exposure period from the medium of concern (e.g. $\mu g/m^3$ [air]; or $\mu g/L$ [water]; or mg/kg [soil])

CR = contact rate, i.e. the amount of contaminated medium contacted per unit time or event (e.g. inhalation rate in m^3/day [air]; or ingestion rate in mg/day [soil], or L/day [water])

CF = conversion factor

FI = fraction of intake from contaminated source (dimensionless)

ABS_f = bioavailability or absorption factor (%)

EF = exposure frequency (days/years)

ED = exposure duration (years)

BW = body weight, i.e. the average body weight over the exposure period (kg)

AT = averaging time (period over which exposure is averaged – days)
   = $ED \times 365$ days/year, for noncarcinogenic effects
   = 70 years $\times$ 365 days/year, for carcinogenic effects (assuming a 70-year average lifetime)

exposure. A maximum daily dose (MDD) will typically be used in estimating acute or subchronic exposures.

6.2.1.3   Toxicity assessment

A toxicity assessment is conducted as part of the human health risk assessment, in order to determine qualitatively and quantitatively the potential adverse health effects that could result from human exposure to environmental contaminants. This involves an evaluation of the types of adverse health effects associated with chemical exposures, the relationship between the magnitude of exposure and adverse effects, and related uncertainties such as the weight-of-evidence of a particular chemical's carcinogenicity in humans.

A comprehensive toxicity assessment for chemicals found at contaminated sites is generally accomplished in two steps: hazard effects assessment and dose–response assessment. Hazard effects assessment is the process used to determine whether exposure to an agent can cause an increase in the incidence of an adverse health effect (e.g. cancer, birth defects, etc.); it involves a characterization of the nature and strength of the evidence of causation. Dose–response assessment is the process of quantitatively evaluating the toxicity information and characterizing the relationship between the dose of the contaminant administered or received (i.e. exposure to an agent) and the incidence of adverse health effects in the exposed populations; it is the process by which the potency of the compounds is estimated by use of dose–response relationships. These steps are discussed in more detail elsewhere in the literature (e.g. Klaassen et al., 1986; USEPA, 1989a).

Evaluation of chemical carcinogenicity   For the purpose of human health risk assessment, chemicals are usually categorized into carcinogenic and noncarcinogenic groups. Chemicals that give rise to toxic endpoints other than cancer and gene mutations are often referred to as 'systemic toxicants' because of their effects on the function of various organ systems; the toxic endpoints are referred to as 'noncancer or systemic toxicity'. Most chemicals that produce noncancer toxicity do not cause a similar degree of toxicity in all organs, but usually demonstrate major toxicity to one or two organs; these are referred to as the target organs of toxicity for the chemicals (Klaassen et al., 1986; USEPA, 1989a). In addition, chemicals that cause cancer and gene mutations also commonly evoke other toxic effects (i.e. systemic toxicity).

Carcinogenic chemicals are generally classified into several categories, depending on the 'weight-of-evidence' or 'strength-of-evidence' available on the particular chemical's carcinogenicity (Hallenbeck and Cunningham, 1988; Huckle, 1991; IARC, 1982; USDHS, 1989; USEPA, 1986). A chemical's potential for human carcinogenicity is inferred from the available information relevant to the potential carcinogenicity of the chemical, and from judgments as to the quality of the available studies. The two evaluation philosophies – one based on 'weight-of-evidence' and the other on 'strength-of-evidence' – have found common acceptance and usage. Systems that employ the weight-of-evidence evaluations consider and balance the negative indicators of carcinogenicity with those showing carcinogenic activity; schemes using the strength-of-evidence evaluations consider combined strengths of all positive animal tests to rank a chemical without evaluating negative studies, nor considering potency or mechanisms (Huckle, 1991).

Noncarcinogens generally are believed to operate by 'threshold' mechanisms, i.e. the manifestation of systemic effects requires a threshold level of exposure or dose to be exceeded during a continuous exposure episode. Thus, noncancer or systemic toxicity is generally treated as if there is an identifiable exposure threshold below which there are no observable adverse effects. This characteristic distinguishes systemic endpoints from carcinogenic and mutagenic endpoints, which are often treated as 'non-threshold' processes. That is, the threshold concept and principle is not applicable for carcinogens, since it is believed that no thresholds exist for this group. It is noteworthy, however, that there is a belief among some toxicologists that certain carcinogens require a threshold exposure level to be exceeded in order to provoke carcinogenic effects.

Determination of toxicity parameters for noncarcinogenic effects Traditionally, risk decisions on systemic toxicity have been made using the concept of 'acceptable daily intake' (ADI), or by using the so-called reference dose (RfD). The ADI is the amount of a chemical (in mg/kg body weight/day) to which a receptor can be exposed on a daily basis over an extended period of time – usually a lifetime – without suffering a deleterious effect. The RfD is defined as the maximum amount of a chemical (in mg/kg body weight/day) that the human body can absorb without experiencing chronic health effects. For exposure of humans to the noncarcinogenic effects of environmental chemicals, the ADI or RfD is used as a measure of exposure considered to be without adverse effects. Although often used interchangeably, RfDs are based on a more rigorously-defined methodology and this parameter is therefore preferred to ADI.

The RfD provides an estimate of the continuous daily exposure of a noncarcinogenic substance for the general human population (including sensitive subgroups) which appears to be without an appreciable risk of deleterious effects. RfDs have been established as thresholds of exposure to toxic substances below which there should be no adverse health impact. These thresholds have been established on a substance-specific basis for oral and inhalation exposures, taking into account evidence from both human epidemiologic and laboratory toxicologic studies.

An elaborate discussion on the derivation of RfDs and ADIs can be found in the literature elsewhere (e.g. Dourson and Stara, 1983; USEPA, 1986, 1989a, 1989c, 1989d).

Determination of toxicity parameters for carcinogenic effects Under the no-threshold assumption, exposure to any level of a carcinogen is considered to have a finite risk of inducing cancer. Assuming a no-threshold situation for carcinogenic effects, an estimate of the excess cancer per unit dose, called the cancer slope factor (SF), is used to develop risk decisions for contaminated site problems. The SF, also called cancer potency factor or potency slope, is a measure of the carcinogenic toxicity or potency of a chemical. It is the cancer risk (proportion affected) per unit of dose (i.e. risk per mg/ kg/day).

In evaluating risks from chemicals found in certain environmental sources, dose–response measures may be expressed as risk per concentration unit. These measures may include the unit cancer risk (UCR) for air (i.e. inhalation UCR) and the unit cancer risk for drinking water (i.e. oral UCR). The continuous lifetime exposure concentration units for air and drinking water are usually expressed in micrograms per cubic meter (μg/m^3) and micrograms per liter (μg/L), respectively.

An elaborate discussion on the derivation of SFs and UCRs can be found in the literature elsewhere (e.g. Dourson and Stara, 1983; USEPA, 1986, 1989a, 1989c).

The use of surrogate toxicity parameters   In general, the toxicity parameters are dependent on the route of exposure. Nonetheless, oral RfDs and SFs are often used for both ingestion and dermal exposures to some chemicals that affect receptors through a systemic action. In fact, in a number of situations, it is appropriate to use oral SFs and RfDs as surrogate values to estimate systemic toxicity as a result of dermal absorption of a chemical (DTSC, 1994; USEPA, 1989d); this will, however, be inappropriate if the chemical affects the receptor contacts through direct local action at the point of application. Also, the use of the oral SF or oral RfD directly does not correct for differences in absorption and metabolism between the oral and dermal routes; typically, absorption fractions of 0.10 and 0.01 are applied to organic and inorganic chemicals, respectively. Furthermore, direct toxic effects on the skin are not accounted for. Thus, the use of an oral SF or oral RfD for the dermal route may result in an over- or underestimation of the true risk or hazard, depending on the nature of the chemicals involved. Consequently, the use of the oral toxicity value as a surrogate for a dermal value will tend to increase the uncertainty in the estimation of risks and hazards; however, this is not generally expected to significantly underestimate the risk or hazard relative to the other routes of exposure that are evaluated in the risk assessment (DTSC, 1994). Similarly, inhalation SFs and RfDs may be used as surrogates for oral (comprising both ingestion and dermal) exposures for those chemicals lacking oral toxicity values, and vice versa.

In other situations, toxicity values to be used in characterizing risks are available only for certain chemicals within a chemical class. In such cases, toxicity parameters may be estimated for chemical groups according to structure–activity relationships or other similarities. Once again, it is expected that significant uncertainties may result by using this type of approach, and such uncertainties should be documented as part of the risk evaluation process.

6.2.1.4   Risk characterization

Risk characterization consists of estimating the probable incidence of adverse impacts to potential receptors under various exposure conditions that are associated with a hazard situation. It involves an integration of the toxicity and exposure assessments, resulting in a quantitative estimation of the actual and potential risks and/or hazards due to exposure to each key chemical constituent, and also the possible additive effects of exposure to mixtures of the chemicals of potential concern. Typically, the risks to potentially exposed populations resulting from exposure to the site contaminants are characterized through a calculation of noncarcinogenic hazard quotients and indices and/or carcinogenic risks (CAPCOA, 1990; CDHS, 1986; USEPA, 1989a). These parameters can then be compared with benchmark standards in order to arrive at risk decisions about a contaminated site.

Aggregate effects of chemical mixtures   The contaminants found at contaminated sites tend to be heterogeneous and variable mixtures that may contain several distinct compounds, distributed over wide spatial regions and across several environmental compartments. The risk assessment process must address the cumulative health risks for the chemical mixtures, to include the multiple endpoints or effects and also the uncertainties in the dose–response functions for each effect. The common method of approach assumes additivity of effects for carcinogens when evaluating chemical mixtures or multiple carcinogens. Prior to the summation of aggregate risks, however, estimated cancer risks should preferably be segregated by 'weight-of-evidence' or 'strength-of-evidence' category for the contaminants at the site, the goal being to provide a clear understanding of the risk contribution of each category of carcinogen. For multiple pollutant exposures to noncarcinogens and noncarcinogenic effects of carcinogens, constituents should be grouped by the same mode of toxicological action (i.e. those which induce the same toxicological endpoint, such as liver toxicity). Cumulative noncarcinogenic risk is evaluated through the use of a hazard index that is generated for each toxicological endpoint. Thus, it often becomes necessary to segregate chemicals by organ-specific toxicity, since strict additivity without consideration for target-organ toxicities could overestimate and overstate potential hazards (USEPA, 1986, 1989a). Consequently, the hazard index is preferably calculated only after putting chemicals into groups with the same physiologic endpoints.

Estimation of carcinogenic risks   For potential carcinogens, risk is defined by the incremental probability of an individual developing cancer over a lifetime as a result of exposure to a carcinogen. Carcinogenic risks can be estimated by combining information about the carcinogenic potency of a chemical and a receptor's exposure to the substance. The carcinogenic effects of site contaminants are typically calculated using the linear low-dose (valid only at low risk levels [i.e. estimated risks <0.01]) and the one-hit (for sites where chemical intakes may be high [i.e. potential risks >0.01]) cancer risk models (USEPA, 1989a), represented by the relationships shown in Box 6.2A and Box 6.2B, respectively. Since the method of approach for assessing the cumulative health risks from chemical mixtures generally assumes additivity of effects for carcinogens when evaluating multiple carcinogens, for multiple carcinogenic chemicals and multiple exposure routes/pathways, the aggregate cancer risk for all exposure pathways and all contaminants associated with a contaminated site can be estimated by the equations shown in Boxes 6.2. The combination of risks across exposure pathways is based on the assumption that the same receptors would consistently experience the reasonable maximum exposure via the multiple pathways. Thus, if specific pathways do not affect the same individual or receptor group, risks should not be combined under those circumstances.

As a rule-of-thumb, incremental risks of between 10^{-4} and 10^{-7} are generally perceived as being reasonable and adequate for the protection of human health and the environment, with 10^{-6} often used as the 'point-of-departure'. In reality, however, populations may be exposed to the same constituents from sources unrelated to a specific project site. Consequently, it is preferred that the estimated carcinogenic risk is well below the 10^{-6} benchmark level, to allow for a reasonable margin of protection for populations potentially at risk.

Estimation of noncarcinogenic risks   The noncancer effects associated with contaminants present at a contaminated site are usually expressed by the hazard quotient (HQ) and/or the hazard index (HI), calculated as the ratio of the estimated chemical exposure level to the route-specific reference dose (USEPA, 1989a). For multiple noncarcinogenic

---

Box 6.2A   The low-dose model for the estimation of low-level carcinogenic risks

$$\text{Total Cancer Risk, } TCR_{\text{lo-risk}} = \sum_{j=1}^{p}\sum_{i=1}^{n}(CDI_{ij} \times SF_{ij})$$

where
TCR = probability of an individual developing cancer (dimensionless)
CDI_{ij} = chronic daily intake for the ith contaminant and jth pathway (mg/kg-day)
SF_{ij} = slope factor for the ith contaminant and jth pathway/exposure route
         ($[\text{mg/kg-day}]^{-1}$)
n = total number of carcinogens
p = total number of pathways or exposure routes

Box 6.2B   The one-hit model for the estimation of high-level carcinogenic risks

$$\text{Total Cancer Risk, } TCR_{\text{hi-risk}} = \sum_{j=1}^{p}\sum_{i=1}^{n}[1 - \exp(-CDI_{ij} \times SF_{ij})]$$

where
TCR = probability of an individual developing cancer (dimensionless)
CDI_{ij} = chronic daily intake for the ith contaminant and jth pathway (mg/kg-day)
SF_{ij} = slope factor for the ith contaminant and jth pathway/exposure route
         ($[\text{mg/kg-day}]^{-1}$)
n = total number of carcinogens
p = total number of pathways or exposure routes

Box 6.3   General equation for the estimation of noncarcinogenic risks

$$\text{Total Hazard Index, } THI = \sum_{j=1}^{p}\sum_{i=1}^{n}\frac{E_{ij}}{RfD_{ij}}$$

$$= \sum_{j=1}^{p}\sum_{i=1}^{n}[HQ]_{ij}$$

where:
E_{ij} = exposure level (or intake) for the ith contaminant and jth pathway (mg/kg-day)
RfD_{ij} = acceptable intake level (or reference dose) for ith contaminant and jth pathway/
          exposure route (mg/kg-day)
$[HQ]_{ij}$ = hazard quotient for ith contaminant and jth pathway
n = total number of chemicals showing noncarcinogenic effects
p = total number of pathways or exposure routes

---

effects of several chemical compounds and multiple exposure routes/pathways, the aggregate noncancer risk for all exposure pathways and all contaminants associated with a contaminated site can be estimated by the equation shown in Box 6.3. Cumulative risk is evaluated through the use of a hazard index that is generated for each health or toxicological 'endpoint', with chemicals having the same endpoint being generally included in a hazard index calculation. The combination of hazard quotients across exposure pathways is based on the assumption that the same receptors would consistently experience the reasonable maximum exposure via the multiple pathways. Thus, if specific pathways do not affect the same individual or receptor group, hazard quotients should not be combined under those circumstances.

As a rule-of-thumb in the interpretation of the results from HI calculations, a reference value of $HI \leqslant 1$ should be taken as the acceptable benchmark. For HI values greater than unity (i.e. $HI > 1$), the higher the value, the greater is the likelihood of adverse noncarcinogenic health impacts. In fact, since populations may be exposed to the same constituents from sources unrelated to a specific site, it is preferred that the estimated noncarcinogenic hazard index be well below the benchmark level of unity, to allow for additional margin of protectiveness for populations potentially at risk. Indeed, if any calculated hazard index exceeds unity, then the health-based criterion for the chemical mixture has been exceeded and the need for interim corrective measures must be considered.

6.2.2 Ecological Risk Assessment

An ecological risk assessment (ERA) involves the qualitative and/or quantitative appraisal of the actual or potential effects of contaminated sites on plants and animals other than humans and domesticated species (USEPA, 1989b, 1990b). In general, a contaminant entering the environment may cause adverse effects only if: (1) the contaminant exists in a form and concentration sufficient to cause harm; (2) the contaminant comes in contact with organisms or environmental media with which it can interact; and (3) the interaction that takes place is detrimental to life functions.

The objectives of an ERA consist of identifying and estimating the potential ecological impacts associated with chemicals emanating from a contaminated site. Ideally, the ERA would estimate the potential for occurrence of adverse effects that are manifested as changes in the diversity, health, and behavior of the constellation of organisms that share a given environment over time. Ecological areas included in an ERA should therefore not be limited by property boundaries of a study area, if affected environments or habitats are located beyond the property boundaries.

The basic components of an ERA program will generally consist of several tasks, including the following:

- Compilation of relevant site data for the study area.
- Determination of background (ambient) concentrations of the site contaminants in the study area.
- Identification of the contaminants of potential ecological concern.
- Identification and location of habitats and environments at the study area and its vicinity.
- Selection of indicator species or habitats.
- Ecotoxicity assessment and/or bioassay of selected indicator species.
- Development of an appropriate conceptual site model (CSM) and/or foodchain diagram.
- Establishment of appropriate assessment endpoints for all chemicals of potential ecological concern.

- Characterization of exposure based on environmental fate and transport of the chemicals of potential ecological concern, as predicted using the CSM and ecological foodchain considerations.
- Development of ecological risk characterization parameters for the indicator species/ habitats.

Where it is deemed necessary, a more detailed assessment that is comprised of a biological diversity analysis and population studies may become part of the overall ERA process. Overall, the process used to evaluate environmental or ecological risks parallels that used in the evaluation of human health risks (as discussed in the preceding section). In both cases, potential risks are determined by the integration of information on chemical exposures with toxicological data for the contaminants of potential concern. Unlike endangerment assessment for human populations, however, ERAs often lack significant amounts of critical and credible data necessary for a comprehensive quantitative evaluation. Nonetheless, the pertinent data requirements should be identified and categorized insofar as practicable, so that reliable risk estimates can be generated.

6.2.2.1   Hazard assessment

In the design of an ERA program for a contaminated site, common elements of populations, communities, and ecosystems should be clearly defined, in order to establish a basis for the development of a logical framework that can be used to characterize risks at the project site. The general elements of the hazard assessment process needed for an ERA are presented below.

Identifying the types of ecosystems, ecological habitats, and community structure   The different types of ecosystems have unique combinations of physical, chemical, and biological characteristics, and thus may respond to contamination in their own unique ways. The physical and chemical structure of an ecosystem may determine how contaminants affect its resident species, and the biological interactions may determine where and how the contaminants move in the environment and which species are exposed to particular concentrations. The general types of ecosystems normally investigated in an ERA fall into two broad categories: terrestrial and aquatic ecosystems, with wetlands serving as a zone of transition between terrestrial and aquatic environments.

In general, the different levels of an ecological community are studied to determine if they exhibit any evidence of stress. If the community appears to have been disturbed, the goal will be to characterize the source(s) of the stress and, specifically, to focus on the degree to which the release of chemical constituents has caused the disturbance or possibly exacerbated an existing problem.

Defining the nature and ecological effects of site contaminants   Although a contaminant may cause illness and/or death to individual organisms, its effects on the structure and function of ecological assemblages or interlinkages may be measured in terms quite different from those used to describe individual effects. Consequently, the hazard assessment should include a wider spectrum of ecological effects on individual organisms as well as the ecological interlinkages. Furthermore, the biological, chemical and environmental factors perceived to influence the ecological effects of contaminants should be identified and succinctly described.

Selection of indicator species   It generally is not feasible to evaluate every species that may be present at a contaminated site and its vicinity. Consequently, selected target or indicator species will normally be chosen for the evaluation of an ERA. By using reasonably conservative assumptions in the overall assessment, it is rationalized that adequate protection of selected indicator species will provide protection for all other significant environmental species as well. It is noteworthy that not every organism may be suitable for use as indicator species in the evaluation of contaminant impacts on ecological systems. Thus, several general considerations and specific criteria should be used to guide the selection of target species in an ERA (USEPA, 1989b, 1990b). For instance, it is important to consider the effects of site contaminants on both an endangered population as well as on the habitats critical to their survival.

Identification of ecological assessment endpoints   The development of an ERA requires the identification of one or several ecological assessment endpoints. These endpoints define the environmental resources which are to be protected, and which, if found to be impacted, determine the need for corrective actions. The selection of appropriate site-specific assessment endpoints is therefore crucial to the development of a cost-effective site characterization and/or site restoration program that is protective of potential ecological receptors.

6.2.2.2 Exposure assessment

The objectives of the exposure assessment are to define contaminant behaviors; identify potential ecological receptors; determine exposure routes by which contaminants may reach ecological receptors; and estimate the degree of contact and/or intakes of the chemicals of concern by the potential receptors. In general, the nature of exposure scenarios determines the potential for any adverse impacts. Hence, a foodchain (also called foodweb) – which gives a simplified and generic conceptual representation of typical interlinkages resulting from the consumption, uptake, and absorption processes associated with an ecological community – is normally constructed for the target species, to facilitate the development of realistic exposure scenarios. Subsequently, the exposure routes are selected based on the behavior patterns and/or ecological niches of the target species and communities.

The amount of a target species' exposure to contamination from a contaminated site is based on the maximum plausible exposure concentrations of the chemicals in the affected environmental matrices. The total daily exposure (in mg/kg-day) of target species can be calculated by summing the amounts of constituents ingested and absorbed from all sources (e.g. soil, vegetation, surface water, fish tissue, and other target species), and also that absorbed through inhalation and dermal contacts. Analytical procedures used to estimate receptor exposures to chemicals in various contaminated media (such as a wildlife or a game's daily chemical exposure and the resulting body burden) are similar to that discussed under human health risk assessment in Section 6.2.1.2. Details of the applicable models can be found elsewhere in the literature of endangerment assessment (e.g. CAPCOA, 1990; Paustenbach, 1988; Suter, 1993).

6.2.2.3   Ecotoxicity assessment

Similar to the human health endangerment assessment (discussed in Section 6.2.1.3), the scientific literature is reviewed to obtain ecotoxicity information for the chemicals of potential ecological concern that are associated with the contaminated site. Additional data may have to be developed via field sampling and analysis and/or bioassays. Subsequently, critical toxicity values for the contaminants of concern are derived for the target ecological receptor species and ecological communities of concern, to be used in characterizing risks associated with the site.

6.2.2.4   Risk characterization

Ecological risk characterization consists of steps similar to that discussed under the human endangerment assessment (Section 6.2.1.4). This entails both temporal and spatial components, requiring an evaluation of the probability or likelihood of an adverse effect occurring; the degree of permanence and/or reversibility of each effect; the magnitude of each effect; and receptor populations or habitats that will be affected. Typically, a quantitative ecological risk characterization is accomplished by using the ecological quotient (EQ) method – similar to the hazard quotient employed in human health risk characterizations. In the EQ approach, the exposure point concentration or estimated daily dose is compared to a benchmark critical toxicity parameter, as follows:

$$\text{Ecological Quotient } (EQ) = \frac{\text{exposure point concentration or estimated daily dose}}{\text{benchmark critical ecotoxicity parameter or a surrogate}}$$

The denominator represents the concentration that produces an assessment endpoint (e.g. toxic effects) in target species. The EQ estimates the risk of a site contaminant to indicator species, independent of the interactions among species or between different chemicals of potential ecological concern. In general, if $EQ \leqslant 1$, then an acceptable risk is indicated. Conversely, an $EQ > 1$ calls for action or further refined investigations, due to the possibility of unacceptable levels of risk to potential ecological receptors.

6.3   CONDUCTING CONTAMINATED SITE RISK ASSESSMENTS

Human populations and ecological receptors are continuously in contact with varying amounts of environmental contaminants present in air, water, soil, and food. Thus, methods for linking contaminant sources in the multiple environmental media to human and ecological receptor exposures are often necessary to facilitate the development of sound site characterization and site restoration programs. Invariably, the methods of approach should, at a minimum, be comprised of the following activities:

- Identification of the sources of contamination.
- Determination of the contaminant migration pathways.
- Identification of populations potentially at risk.
- Determination of the specific chemicals of potential concern.
- Determination of frequency of potential receptor exposures to site contaminants.
- Evaluation of contaminant exposure levels.

- Determination of receptor response to chemical exposures.
- Estimation of impacts or damage resulting from receptor exposures to the chemicals of potential concern.

Ultimately, potential risks are estimated by considering the probability or likelihood of occurrence of harm; the intrinsic harmful features or properties of specified hazards; the populations potentially at risk; the exposure scenarios; and the extent of expected harm and potential effects. The results of the risk assessment are typically used as follows:

- To document the magnitude of risk at a site, including the primary causes of the risk.
- To facilitate a determination as to whether further response action is necessary at a site, or to support and justify a 'no-further-action' decision, based on both existing and anticipated exposure scenarios associated with the site.
- To prioritize the need for site restoration, and provide a basis for quantifying remedial action objectives.

Traditionally, contaminated site endangerment assessments have focused almost exclusively on risks to human health, often ignoring potential ecological effects because of the common but mistaken belief that protection of human health automatically protects non-human organisms (Suter, 1993). In fact, it is true that human health risks in most situations are more substantial than ecological risks, and mitigative actions taken to alleviate risks to human health are often sufficient to mitigate potential ecological risks at the same time. However, in some other situations non-human organisms, populations, or ecosystems may be more sensitive to site contaminants than are human receptors. Consequently, ecological risk assessment programs should be considered as an equally important component of the management of contaminated site problems, in order to arrive at site restoration decisions that offer an adequate level of protection for both human and ecological populations that are potentially at risk.

6.4 ATTRIBUTES OF CONTAMINATED SITE RISK ASSESSMENT

The risk assessment process will generally utilize the best available scientific knowledge and data to establish case-specific responses to contaminated site management problems. In particular, the assessment of human health and environmental risks associated with potentially contaminated sites may contribute, in a significant way, to the processes involved in site characterization programs; in corrective action planning; in risk mitigation and risk management strategies; and in the overall management of problems associated with potentially contaminated sites.

Depending on the scope of the analysis, methods used in estimating risks may be either qualitative or quantitative. Thus, the process may be one of data analysis or modeling, or a combination of the two. In fact, the process of quantifying risks does, by its very nature, give a better understanding of the strengths and weaknesses of the potential hazards being examined. It shows where a given effort can do the most good in modifying a system in order to improve its safety and efficiency.

The major attributes of risk assessment that are relevant to the management of contaminated site problems include the following:

- Identification and ranking of existing and anticipated potential hazards.
- Explicit consideration of current and possible future exposure scenarios.
- Qualification and/or quantification of risks associated with the full range of hazard situations, system responses, and exposure scenarios.
- Identification and analysis of the sources of uncertainties associated with site characterization programs.
- Determination of cost-effective risk reduction policies, via an evaluation of risk-based remedial alternatives and/or the adoption of efficient risk management and risk prevention programs.

Each attribute will ultimately play an important role in the overall site characterization and corrective action response programs.

In general, data generated in a risk assessment are used to determine the need for, and the degree of remediation required for potentially contaminated sites. It is noteworthy, however, that there are inherent uncertainties associated with risk assessments due to the fact that the risk assessor's knowledge of the causative events and controlling factors usually is limited, and also because the results obtained depend, to a reasonable extent, on the methodology and assumptions used. Furthermore, risk assessment can impose potential delays in the implementation of corrective action programs; however, the overall gain in program efficiency is likely to more than compensate for the delays.

6.5 REFERENCES

Asante-Duah, D.K. (1990). Quantitative risk assessment as a decision tool for hazardous waste management. In: *Proceedings of 44-th Purdue Industrial Waste Conference (May 1989), pp. 111–123*. Lewis Publishers, Inc., Chelsea, Michigan.

Asante-Duah, D.K. (1993). *Hazardous Waste Risk Assessment*. CRC Press/Lewis Publishers, Inc., Boca Raton, Florida.

Bowles, D.S., L.R. Anderson, and T.F. Glover (1987). Design level risk assessment for dams. In: *Proceedings of Structural Congress, ASCE*, pp. 210–225. Florida.

CAPCOA (California Air Pollution Control Officers Association) (1990). *Air Toxics 'Hot Spots' Program. Risk Assessment Guidelines*. California Air Pollution Control Officers Association, California.

CDHS (California Department of Health Services) (1986). *The California Site Mitigation Decision Tree Manual*. California Department of Health Services, Toxic Substances Control Division, Sacramento, California.

DOE (US Department of Energy) (1987). *The Remedial Action Priority System (RAPS): Mathematical Formulations*. US Dept. of Energy, Office of Environ., Safety & Health, Washington, DC.

Dourson, M.L. and J.F. Stara (1983). Regulatory history and experimental support of uncertainty (safety) factors. *Reg. Toxicol. Pharmacol.* 3, 224–238.

DTSC (Department of Toxic Substances Control) (1994). *Preliminary Endangered Assessment Guidance Manual* (A guidance manual for evaluating hazardous substance release sites). California Environmental Protection Agency, DTSC, Sacramento, California.

Hallenbeck, W.H. and K.M. Cunningham (1988). *Quantitative Risk Assessment for Environmental and Occupational Health*, 4th Printing. Lewis Publishers, Inc., Chelsea, Michigan.

Huckle, K.R. (1991). *Risk Assessment – Regulatory Need or Nightmare*. Shell Publications, Shell Centre, London, England, 8 pp.

IARC (International Agency for Research on Cancer) (1982). *IARC Monographs on the Evaluation of the Carcinogenic Risk of Chemicals to Humans. Chemicals, Industrial Processes and Industries Associated with Cancer in Humans.* Supplement 4. IARC, Lyons, France, 292 pp.

Klaassen, C.D., M.O. Amdur, and J. Doull (eds) (1986). *Casarett and Doull's Toxicology: The Basic Science of Poisons,* 3rd edn. Macmillan Publishing Company, New York.

NRC (National Research Council) (1983). *Risk Assessment in the Federal Government: Managing the Process.* National Academy Press, Washington, DC.

OSA (Office of Scientific Affairs) (1993). *Supplemental Guidance for Human Health Multimedia Risk Assessments of Hazardous Waste Sites and Permitted Facilities.* Cal EPA, DTSC, Sacramento, California.

Paustenbach, D.J. (ed.) (1988). *The Risk Assessment of Environmental Hazards: A Textbook of Case Studies.* John Wiley & Sons, New York.

Rowe, W.D. (1977). *An Anatomy of Risk.* John Wiley & Sons, New York.

Suter II, G.W. (1993). *Ecological Risk Assessment.* Lewis Publishers, Chelsea, Michigan.

USDHS (US Department of Health and Human Services) (1989). *Public Health Service.* Fifth Annual Report on Carcinogens. Summary.

USEPA (US Environmental Protection Agency) (1984). *Risk Assessment and Management: Framework for Decision Making.* EPA 600/9-85-002, Washington, DC.

USEPA (US Environmental Protection Agency) (1986). Guidelines for the health risk assessment of chemical mixtures. *Federal Register* **51**(185), 34014–34025, CFR 2984, 24 September.

USEPA (US Environmental Protection Agency) (1987). *RCRA Facility Investigation (RFI) Guidance.* EPA/530/SW-87/001, Washington, DC.

USEPA (US Environmental Protection Agency) (1988). *Superfund Exposure Assessment Manual,* Report No. EPA/540/1-88/001, OSWER Directive 9285.5-1, USEPA, Office of Remedial Response, Washington, DC.

USEPA (US Environmental Protection Agency) (1989a). *Risk Assessment Guidance for Superfund.* Vol. I – *Human Health Evaluation Manual* (Part A). EPA/540/1-89/002. Office of Emergency and Remedial Response, Washington, DC.

USEPA (US Environmental Protection Agency) (1989b). *Risk Assessment Guidance for Superfund.* Vol. II – *Environmental Evaluation Manual.* EPA/540/1-89/001. Office of Emergency and Remedial Response, Washington, DC.

USEPA (US Environmental Protection Agency) (1989c). *Exposure Factors Handbook.* EPA/600/8-89/043. Office of Health and Environmental Assessment, Washington, DC.

USEPA (US Environmental Protection Agency) (1989d). *Interim Methods for Development of Inhalation Reference Doses.* EPA/600/8-88/066F. Office of Health and Environmental Assessment, Washington, DC.

USEPA (US Environmental Protection Agency) (1990a). *Guidance for Data Usability in Risk Assessment.* Interim Final. EPA/540/G-90/008. Office of Emergency and Remedial Response, Washington, DC.

USEPA (US Environmental Protection Agency) (1990b). *State of the Practice of Ecological Risk Assessment Document.* Office of Pesticides and Toxic Substances, USEPA draft report. Washington, DC.

USEPA (US Environmental Protection Agency) (1991). *Risk Assessment Guidance for Superfund.* Vol. I – *Human Health Evaluation Manual.* Supplemental Guidance. Standard Default Exposure Factors (Interim Final). March 1991. Office of Emergency and Remedial Response, Washington, DC. OSWER Directive: 9285.6-03.

PART III
DEVELOPMENT OF SITE RESTORATION

Chapter Seven

Development of risk-based site restoration goals

An important consideration in the development of site restoration programs for contaminated sites is the level of cleanup to be achieved during possible remedial activities. The site cleanup level is a site-specific criterion that a remedial action would have to satisfy in order to keep exposures of potential receptors to contaminant levels at an 'acceptable level'. The acceptable level corresponds to a contaminant concentration in specific environmental media which, when exceeded, may result in significant risks of adverse impact to potential receptors. This target level generally tends to drive the cleanup process for a contaminated site, and therefore represents the maximum acceptable contaminant level for site restoration decisions.

Oftentimes, pre-existing 'environmental quality criteria' for various environmental media or matrices serve as benchmarks in the assessment of the degree of contamination, and also in the determination of the level of cleanup necessary to protect human health and the environment. An even better and more sophisticated approach involves the use of risk assessment principles to establish site-specific cleanup objectives for contaminated sites that require remediation. Typically, cleanup criteria are developed by 'back-modeling' from a benchmark risk level in order to obtain an acceptable risk-based concentration, or the maximum acceptable concentration. In general, the type of exposure scenarios envisioned and the exposure assumptions used may drive the level of cleanup warranted. Ultimately, the use of such an approach aids in the selection of appropriate remediation options capable of achieving a set of performance goals.

This chapter elaborates a number of analytical relationships that can be adopted or used to estimate contaminated site cleanup levels necessary for site restoration decisions.

7.1 FACTORS AFFECTING THE DEVELOPMENT OF RISK-BASED REMEDIATION GOALS

Due to the possibility for different cleanup levels to be imposed on similar sites potentially contaminated to the same degree, it is important that a systematic approach is used in the development of site-specific cleanup criteria for contaminated sites. It is also important that the determination of the extent of cleanup required at a contaminated site is based on an assessment of the potential risks to both human health and the environment. A number of exposure- and technical-related factors will generally affect the development of remediation objectives and cleanup goals. Several specific factors that are important to the process of establishing contaminated site cleanup criteria relate to the following:

- Nature and level of risks involved
- Regulatory requirements and/or guidelines
- Migration and exposure pathways (from contaminant sources to receptors)
- Individual site characteristics affecting exposure
- Current and future beneficial uses of the affected land and subsurface resources
- Variability in exposure scenarios
- Probability of occurrence of exposure to the populations potentially at risk
- Possibilities of receptor exposures to elevated levels of other contamination not related to site activities
- Sensitivity and vulnerability of the populations potentially at risk
- Potential effects of site contamination on human and ecological receptors
- Reliability of scientific data relating to exposure assessment, toxicity data, and risk models.

The type of exposure scenarios envisioned for a contaminated site and its vicinity usually will significantly affect whatever is considered to be an acceptable cleanup level. Thus, entirely different cleanup levels may be needed for similar pieces of equally contaminated sites, based on the differences in the exposure scenarios. That is, the same amount of contamination at similar sites does not necessarily call for the same level of cleanup. In general, however, the cleanup must attain contaminant levels that are protective of all receptors for both current and future land-uses. After defining the critical pathways and exposure scenarios associated with a project site, it often is possible to calculate the various media concentrations at or below which potential receptor exposures will pose no significant risks to the exposed populations.

7.1.1 Requirements and Criteria for Establishing Risk-based Cleanup Levels

Risk-based cleanup levels (*RBCLs*) may be established for various environmental matrices at a contaminated site by manipulating the risk and exposure models previously presented in Chapter 6. *RCBLs* are typically established for both carcinogenic and noncarcinogenic effects of the site contaminants, with the more stringent criteria usually being selected as a site restoration goal; invariably, the carcinogenic *RBCL* tends to be more stringent in most situations where both values exist. In addition, the following criteria, assuming dose additivity, must be met by the site restoration goal:

$$\sum_{j=1}^{p} \sum_{i=1}^{n} \frac{CMAX_{ij}}{RBCL_{ij}} < 1$$

where $CMAX_{ij}$ is the prevailing maximum concentration of contaminant i in environmental matrix j, and $RBCL_{ij}$ is the risk-based cleanup level for contaminant i in medium j. When these conditions are satisfied, then the *RBCL* represents a maximum acceptable contaminant level for site cleanup that will be protective of public health and the environment. Thus, exceeding the *RBCL* will usually call for the development and implementation of a site remediation program.

For most contaminated site problems, soils and groundwater apparently represent the most significant media of concern, considering their importance in site characterization and site restoration decisions. The computational procedures for developing *RBCLs* for these environmental media are therefore elaborated below. The same principles can be extended in the formulation and development of *RBCLs* for a variety of other environmental matrices.

7.2 DEVELOPMENT OF RISK-BASED SOIL CLEANUP LEVELS

To determine the *RBCLs* for a chemical compound present in soils at a contaminated site, algebraic manipulations of the hazard index or carcinogenic risk equation and the exposure estimation equations discussed in Chapter 6 can be used to arrive at the appropriate analytical relationships that define the risk-based soil cleanup criteria necessary for a site restoration program. The step-wise computational process involves performing back-calculations from the risk and exposure models in order to arrive at an acceptable soil concentration that is based on health-protective exposure parameters. For chemicals with carcinogenic effects, a target cancer risk of 10^{-6} is typically used in the back-calculation process; a target hazard index of 1.0 is generally used for noncarcinogenic effects.

7.2.1 Soil Cleanup Level for Carcinogenic Contaminants

Box 7.1 contains the general equation for calculating the risk-based site restoration criteria for a single carcinogenic chemical present in soils at a contaminated site. This has been derived by back-calculating from the risk and chemical exposure equations associated with the inhalation of soil emissions, ingestion of soils, and dermal contact with soils.

In a simplified example of the application of this equation (for calculating media-specific *RBCL* for a carcinogenic chemical), consider a hypothetical site located within a residential setting where children may be exposed to site contamination during recreational activities. It has been determined that soil at this playground for young children in the neighborhood is contaminated with methylene chloride. It is expected that children aged 1 to 6 years could be ingesting up to 200 mg of contaminated soil per day during outdoor activities at the impacted playground. The soil *RBCL* associated with the *ingestion only exposure* of 200 mg of soil (contaminated with methylene chloride which has an oral *SF* of 7.5×10^{-3} [mg/kg-day]$^{-1}$) on a daily basis by a 16-kg child over a 5-year exposure period is conservatively estimated to be:

$$RBCL_{mc} = \frac{[10^{-6} \times 16 \times 70 \times 365]}{[0.0075 \times 200 \times 1 \times 1 \times 365 \times 5 \times 10^{-6}]} \approx 149 \, mg/kg$$

The allowable exposure concentration (represented by the soil *RBCL*) for methylene chloride in soils within this residential setting, assuming a benchmark excess lifetime cancer risk level of 10^{-6}, is estimated to be approximately 149 mg/kg. Thus, if

---

Box 7.1   General equation for calculating risk-based soil cleanup level for a carcinogenic
chemical contaminant

$RBCL_{c\text{-soil}} =$

$$\frac{(TCR) \times (BW \times AT \times 365)}{(EF \times ED \times CF) \times \{[SF_i \times IR \times RR \times ABS_a \times AEF \times CF_a] + SF_o[(SIR \times FI \times ABS_{si}) + (SA \times AF \times ABS_{sd} \times SM)]\}}$$

where:

$RBCL_{c\text{-soil}}$	= Acceptable risk-based cleanup level of carcinogenic contaminant in soil (mg/kg)
TCR	= Target cancer risk, usually set at 10^{-6} (dimensionless)
SF_i	= Inhalation slope factor ($[\text{mg/kg-day}]^{-1}$)
SF_o	= Oral slope factor ($[\text{mg/kg-day}]^{-1}$)
IR	= Inhalation rate (m³/day)
RR	= Retention rate of inhaled air (%)
ABS_a	= Percent chemical absorbed into bloodstream (%)
AEF	= Air emissions factor, i.e. PM_{10} particulate emissions or volatilization (kg/m³)
CF_a	= Conversion factor for air emission term (10^6)
SIR	= Soil ingestion rate (mg/day)
CF	= Conversion factor (10^{-6} kg/mg)
FI	= Fraction ingested from contaminated source (dimensionless)
ABS_{si}	= Bioavailability absorption factor for ingestion exposure (%)
ABS_{sd}	= Bioavailability absorption factor for dermal exposures (%)
SA	= Skin surface area available for contact, i.e. surface area of exposed skin (cm²/ event)
AF	= Soil to skin adherence factor, i.e. soil loading on skin (mg/cm²)
SM	= Factor for soil matrix effects (%)
EF	= Exposure frequency (days/years)
ED	= Exposure duration (years)
BW	= Body weight (kg)
AT	= Averaging time (i.e. period over which exposure is averaged) (years)

---

environmental sampling and analysis indicate contamination levels in excess of 149 mg/
kg at this residential playground, then immediate corrective action (such as restricting
access to the playground as an interim measure) should be implemented. It is
noteworthy that other potentially significant exposure routes (e.g. dermal contact and
inhalation) as well as other sources of exposure (e.g. via drinking water and food) have
not been accounted for in this illustrative example, which may further require the need
to lower the calculated *RBCL* for any site restoration decisions.

7.2.2   Soil Cleanup Criteria for the Noncarcinogenic Effects of Site Contaminants

Box 7.2 contains the general equation for calculating the risk-based site restoration
criteria for the noncarcinogenic effects of a single chemical constituent found in soils at
a contaminated site. This has been derived by back-calculating from the hazard and
chemical exposure equations associated with the inhalation of soil emissions, ingestion
of soils, and dermal contact with soils.

In a simplified example of the application of this equation (for calculating media-
specific *RBCL* for the noncarcinogenic effects of a chemical constituent), consider a
hypothetical site located within a residential setting where children may be exposed to
site contamination during recreational activities. It has been determined that soil at this
playground for young children in the neighborhood is contaminated with ethylbenzene.

Box 7.2   General equation for calculating risk-based soil cleanup level for the noncarcinogenic effects of a chemical contaminant

$RBCL_{\text{nc-soil}}$

$$= \frac{(THQ) \times (BW \times AT \times 365)}{(EF \times ED \times CF) \times \left\{ \left[\frac{IR \times RR \times ABS_a}{RfD_i} \times AEF \times CF_a \right] + \frac{1}{RfD_o} \left[(SIR \times FI \times ABS_{si}) + (SA \times AF \times ABS_{sd} \times SM) \right] \right\}}$$

where:

$RBCL_{\text{nc-soil}}$	= Acceptable risk-based cleanup level of noncarcinogenic contaminant in soil (mg/kg)
THQ	= Target hazard quotient (usually equal to 1)
RfD_i	= Inhalation reference dose (mg/kg-day)
RfD_o	= Oral reference dose (mg/kg-day)
IR	= Inhalation rate (m^3/day)
RR	= Retention rate of inhaled air (%)
ABS_a	= Percent chemical absorbed into bloodstream (%)
AEF	= Air emission factor, i.e. PM$_{10}$ particulate emissions or volatilization (kg/m^3)
CF_a	= Conversion factor for air emission term (10^6)
SIR	= Soil ingestion rate (mg/day)
CF	= Conversion factor (10^{-6} kg/mg)
FI	= Fraction ingested from contaminated source (dimensionless)
ABS_{si}	= Bioavailability absorption factor for ingestion exposure (%)
ABS_{sd}	= Bioavailability absorption factor for dermal exposures (%)
SA	= Skin surface area available for contact, i.e. surface area of exposed skin (cm^2/event)
AF	= Soil to skin adherence factor, i.e. soil loading on skin (mg/cm^2)
SM	= Factor for soil matrix effects (%)
EF	= Exposure frequency (days/years)
ED	= Exposure duration (years)
BW	= Body weight (kg)
AT	= Averaging time (i.e. period over which exposure is averaged) (years)

It is expected that children aged 1 to 6 years could be ingesting up to 200 mg of contaminated soil per day during outdoor activities at the impacted playground. The soil *RBCL* associated with the *ingestion only exposure* of 200 mg of soil (contaminated with ethylbenzene which has an oral RfD of 0.1 mg/kg-day) on a daily basis by a 16-kg child over a 5-year exposure period is conservatively estimated to be:

$$RBCL_{\text{ebz}} = \frac{0.1 \times [1 \times 16 \times 5 \times 365]}{[200 \times 1 \times 1 \times 365 \times 5 \times 10^{-6}]} \approx 8000 \text{ mg/kg}$$

The allowable exposure concentration (represented by the soil RBCL) for ethylbenzene in soils within this residential setting is estimated to be approximately 8000 mg/kg. Thus, if environmental sampling and analysis indicate contamination levels in excess of 8000 mg/kg at this residential playground, then immediate corrective action (such as restricting access to the playground as an interim measure) should be implemented. It is noteworthy that other potentially significant exposure routes (e.g. dermal contact and inhalation) as well as other sources of exposure (e.g. via drinking water and food) have not been accounted for in this illustrative example, which may further require the need to lower the calculated *RBCL* for any site restoration decisions.

Box 7.3   General equation for calculating risk-based water cleanup level for a carcinogenic chemical contaminant

$RBCL_{\text{c-water}}$

$$= \frac{TCR \times (BW \times AT \times 365)}{(EF \times ED) \times \{[SF_i \times IR_w \times RR \times ABS_a \times CF_a] + SF_o[(WIR \times FI \times ABS_{si}) + (SA \times K_p \times ET \times ABS_{sd} \times CF)]\}}$$

where:

$RBCL_{\text{c-water}}$	= Acceptable risk-based cleanup level of carcinogenic contaminant in water (mg/L)
TCR	= Target cancer risk, usually set at 10^{-6} (dimensionless)
SF_i	= Inhalation slope factor ($[\text{mg/kg-day}]^{-1}$)
SF_o	= Oral slope factor ($[\text{mg/kg-day}]^{-1}$)
IR_w	= Intake from the inhalation of volatiles (sometimes equivalent to the amount of ingested water) (m³/day)
RR	= Retention rate of inhaled air (%)
ABS_a	= Percent chemical absorbed into bloodstream (%)
CF_a	= Conversion factor for volatiles inhalation term ($1000\,\text{L}/1\,\text{m}^3 = 10^3\,\text{L/m}^3$)
WIR	= Water ingestion rate (L/day)
CF	= Conversion factor ($1\,\text{L}/1000\,\text{cm}^3 = 10^{-3}\,\text{L/cm}^3$)
FI	= Fraction ingested from contaminated source (unitless)
ABS_{si}	= Bioavailability absorption factor for ingestion exposure (%)
ABS_{sd}	= Bioavailability absorption factor for dermal exposures (%)
SA	= Skin surface area available for contact, i.e. surface area of exposed skin (cm²/event)
K_p	= Chemical-specific dermal permeability coefficient from water (cm²/hour)
ET	= Exposure time during water contacts (e.g. during showering/bathing activity) (hours/day)
EF	= Exposure frequency (days/years)
ED	= Exposure duration (years)
BW	= Body weight (kg)
AT	= Averaging time (i.e. period over which exposure is averaged) (years)

7.3   DEVELOPMENT OF RISK-BASED WATER CLEANUP LEVELS

To determine the *RBCLs* for a chemical compound present in water at a contaminated site, algebraic manipulations of the hazard index or carcinogenic risk equation and the exposure estimation equations discussed in Chapter 6 can be used to arrive at the appropriate analytical relationships that define the risk-based water cleanup criteria necessary for a site restoration program. The step-wise computational process involves performing back-calculations from the risk and exposure models in order to arrive at an acceptable water concentration that is based on health-protective exposure parameters. For chemicals with carcinogenic effects, a target risk of 1×10^{-6} is typically used in the back-calculation process; a target hazard index of 1.0 is generally used for noncarcinogenic effects.

7.3.1   Water Cleanup Level for Carcinogenic Contaminants

Box 7.3 contains the relevant equation to use in the development of risk-based site restoration criteria for a single carcinogenic constituent present in water at a contaminated site. This has been derived by back-calculating from the risk and chemical exposure equations associated with the inhalation of contaminants in water (for volatile constituents only), ingestion of water, and dermal contact with water.

In a simplified example of the application of this equation (for calculating media-specific *RBCL* for a carcinogenic chemical), consider the case of a contaminated site that is impacting an underlying water supply aquifer due to contaminant migration into groundwater. This groundwater resource is used for culinary water supply purposes. The water *RBCL* associated with the daily ingestion of 2 liters of water (contaminated with methylene chloride which has an oral *SF* of 7.5×10^{-3} [mg/kg-day]$^{-1}$) by a 70-kg adult over a 70-year lifetime is given by the following approximation:

$$RBCL_{mc} = \frac{[10^{-6} \times 70 \times 70 \times 365]}{[0.0075 \times 2 \times 1 \times 1 \times 365 \times 70]} \approx 0.005 \, \text{mg/L} = 5 \, \mu\text{g/L}$$

Thus, the allowable exposure concentration (represented by the water *RBCL*) for methylene chloride, assuming a benchmark excess lifetime cancer risk level of 10^{-6} is estimated to be $5\,\mu$g/L. Obviously, the inclusion of other pertinent exposure routes (such as inhalation of vapors and also dermal contacts during showering/bathing and washing activities) will likely result in the need for a lower *RBCL* in any site restoration decision.

7.3.2 Water Cleanup Level for the Noncarcinogenic Effects of Site Contaminants

Box 7.4 contains the relevant equation to use in the development of risk-based site restoration criteria for a single noncarcinogenic constituent present in water at a contaminated site. This has been derived by back-calculating from the hazard and chemical exposure equations associated with the inhalation of contaminants in water (for volatile constituents only), ingestion of water, and dermal contact with water.

In a simplified example of the application of this equation (for calculating media-specific *RBCL* for the noncarcinogenic effects of a chemical constituent), consider the case of a contaminated site impacting a multipurpose groundwater supply source due to contaminant migration into an underlying aquifer. This groundwater resource is used for culinary water supply purposes. The water *RBCL* associated with the daily ingestion of 2 liters/day of water (contaminated with ethylbenzene which has an oral RfD of 0.1 mg/kg-day) by a 70-kg adult is approximated by:

$$C_{ebz} = \frac{[0.1 \times 1 \times 70 \times 70 \times 365]}{[2 \times 1 \times 1 \times 365 \times 70]} \approx 3500 \, \mu\text{g/L}$$

Thus, the allowable exposure concentration (represented by the water *RBCL*) for ethylbenzene is estimated to be $3500\,\mu$g/L. Of course, additional exposures via inhalation and dermal contacts during showering/bathing and washing activities may also have to be incorporated to yield an even lower *RBCL*, in order to arrive at a more responsible site restoration decision.

Box 7.4   General equation for calculating risk-based water cleanup level for noncarcinogenic effects of a chemical contaminant

$RBCL_{\text{nc-water}}$

$$= \frac{THQ \times (BW \times AT \times 365)}{(EF \times ED) \times \left\{ \left[\frac{IR_w \times RR \times ABS_a \times CF_a}{RfD_i} \right] + \frac{1}{RfD_o} \left[(WIR \times FI \times ABS_{si}) + (SA \times K_p \times ET \times ABS_{sd} \times CF) \right] \right\}}$$

where:

$RBCL_{\text{nc-water}}$ = Acceptable risk-based cleanup level of noncarcinogenic contaminant in soil (mg/L)

THQ = Target hazard quotient (usually equal to 1)

RfD_i = Inhalation reference dose (mg/kg-day)

RfD_o = Oral reference dose (mg/kg-day)

IR_w = Inhalation intake rate (m³/day)

RR = Retention rate of inhaled air (%)

ABS_a = Percent chemical absorbed into bloodstream (%)

CF_a = Conversion factor for volatiles inhalation term ($1000\,L/1\,m^3 = 10^3\,L/m^3$)

WIR = Water intake rate (L/day)

CF = Conversion factor ($1\,L/1000\,cm^3 = 10^{-3}\,L/cm^3$)

FI = Fraction ingested from contaminated source (dimensionless)

ABS_{si} = Bioavailability absorption factor for ingestion exposure (%)

ABS_{sd} = Bioavailability absorption factor for dermal exposures (%)

SA = Skin surface area available for contact, i.e. surface area of exposed skin (cm²/event)

K_p = Chemical-specific dermal permeability coefficient from water (cm²/hour)

ET = Exposure time during water contacts (e.g. during showering/bathing activity) (hours/day)

EF = Exposure frequency (days/years)

ED = Exposure duration (years)

BW = Body weight (kg)

AT = Averaging time (i.e. period over which exposure is averaged) (years)

7.4   THE SITE RESTORATION DECISION PROCESS

Oftentimes, pre-established environmental quality criteria (EQC) are used to define cleanup goals if they represent an 'acceptable level' with respect to site-specific factors (including the physical setting of the site). However, such EQC may not always be available or they may not be adequate if the presence of multiple contaminants, multiple pathways, or other extraneous factors result in an 'unacceptable' aggregate risk for the site-specific circumstances. In such situations, the appropriate level of protection is determined using alternative site restoration decision strategies. *RBCLs* derived for the various pathways from elaborately defined exposure scenarios will typically aid in developing the site restoration decisions, such that public health and/or the environment are not jeopardized by any residual contamination. The following guidelines may be used to facilitate the process of establishing the media-specific *RBCLs*:

- In developing cleanup criteria, it usually is necessary to establish a benchmark level of risk for the contaminants of concern. Media cleanup standards are generally established within the risk range of 10^{-7} to 10^{-4} (with a lifetime excess cancer risk of 10^{-6} normally used as a point-of-departure) and a hazard index of 1. In any case, it is preferred that the cumulative risk posed by multiple contaminants does not exceed a 10^{-4} cancer risk and/or a hazard index of unity.

- Sensitive ecosystems and habitats or threatened or endangered species may require more stringent standards for their protection.
- If nearby populations are exposed to hazardous constituents from other sources, lower cleanup levels may generally be required than would ordinarily be necessary.
- If exposures to certain hazardous constituents occur through multiple pathways, lower cleanup levels should generally be prescribed.

Ultimately, the scale and urgency of response actions at contaminated sites depend on the degree to which contaminant levels exceed their respective benchmarks or risk-based criteria. Where site remediation is not feasible, the EQC or risk-based criteria can be used to guide land-use restrictions or other forms of risk management actions that are protective of human health and the environment.

In general, the use of risk-based cleanup levels is likely to result in timely, cost-effective, and adequate site restoration programs. As a rule-of-thumb, remedies whose cumulative effects fall within the risk range of approximately 10^{-4} to 10^{-7} for carcinogens, or meet an acceptable hazard level of unity for noncarcinogenic effects, are generally considered protective of human health. Where necessary, however, the potential ecological impacts should also be determined before a final site restoration decision is made. In fact, media cleanup goals should generally be established at contaminant levels protective of both human health and the environment. Oftentimes, however, cleanup levels established for the protection of human health will also be protective of the environment at the same time. But there may be instances where adverse environmental effects may occur at or below contaminant levels that adequately protect human health. Consequently, sensitive ecosystems as well as threatened and endangered species or habitats that may be affected by releases of hazardous contaminants or constituents should, insofar as possible, be evaluated separately as part of the process used to establish media cleanup criteria needed for site restoration initiatives.

7.4.1 A Recommended Health-protective Site Cleanup Level

Oftentimes, the preliminary remediation goal that has been established based on an acceptable risk level or hazard index is for a single contaminant in one environmental matrix. Therefore the risk and hazard associated with multiple contaminants in a multimedia setting are not fully accounted for during the back-modeling process used to establish the *RBCL*s. In contrast, the evaluation of risks associated with a given contaminated site problem usually involves a set of equations designed to estimate hazard and risk for several contaminants present at the site, and for a multiplicity of exposure pathways. Under this latter type of scenario, the computed 'residual' risks that could remain after site remediation are likely to exceed acceptable health-protective limits; consequently, it becomes necessary to establish a modified *RBCL* for the remedy selection process. To obtain the modified *RBCL*, the 'acceptable' contaminant level is estimated in the same way as previously elaborated (Sections 7.2 and 7.3), but with the cumulative effects of multiple contaminants being taken into account through a process of apportioning target risks and hazards among all chemicals of potential concern.

The health-protective cleanup level for carcinogenic chemicals   The acceptable risk level may be apportioned between the chemical constituents contributing to the overall target risk by assuming that each constituent contributes equally or proportionately to the total acceptable risk. The 'risk fraction' obtained for each constituent can then be used to derive the modified *RBCL* by working from the relationships established previously for the computation of *RBCL*s (Sections 7.2 and 7.3). Such an approach will ensure that the sum of all risks from the chemicals involved over all exposure pathways is less than or equal to the set target risk (i.e. $\leqslant 10^{-6}$, as an example).

The health-protective cleanup level for noncarcinogenic constituents   The acceptable hazard level may be apportioned between the chemical constituents contributing to the overall hazard index, assuming each constituent contributes equally or proportionately to the total acceptable hazard index. The 'hazard fraction' obtained for each constituent can then be used to derive the modified *RBCL* by working from the relationships established previously for the computation of *RBCL*s (Sections 7.2 and 7.3). For noncarcinogenic effects of chemicals with the same physiologic endpoint, such an approach will ensure that the sum of all hazard quotients over all exposure pathways for chemicals (with the same toxicological endpoints) is less than or equal to the hazard index criterion of 1.0.

7.4.2   Justification for the Use of Risk-based Cleanup Levels in Site Restoration Decisions

Some important advantages for using *RBCL*s in site restoration decisions include the following (Lesage and Jackson, 1992; Pratt, 1993):

- The cleanup levels defined are based on site-specific data and conditions, which will likely have significant impacts on cleanup costs. This is because the *RBCL*s are specific to the proposed land-use, or are set to ensure that a range of end-uses can be achieved without unacceptable risk to future land-uses.
- The levels defined are sensitive to human health and environmental effects, without necessarily being sensitive to regulatory changes.
- The methodology allows for the derivation of an *asymptotic level* of the contaminants of concern in the impacted media, which represents the cleanup level corresponding to a point of diminishing returns (i.e. the point when monitoring indicates that little additional progress can be made in reducing the contaminant levels). This represents the attainment of contaminant levels below which continued remediation produces negligible reductions in contaminant levels.

The use of site-specific *RBCL*s will normally result in significant cost-savings, because this allows a site management team to employ cost-effective site restoration strategies to achieve significant risk reduction for the particular situation. In fact, used as the cleanup criteria, the *RBCL* could become the driving force behind site remediation costs. It is therefore prudent to allocate adequate resources to develop contaminated site cleanup criteria. The site-specific *RBCL* should generally facilitate the processes involved in the selection and design of cost-effective remedial action alternatives.

7.5 REFERENCES

Lesage, S. and R.E. Jackson (eds) (1992). *Groundwater Contamination and Analysis at Hazardous Waste Sites.* Marcel Dekker, New York.
Pratt, M. (ed.) (1993). *Remedial Processes for Contaminated Land.* Institution of Chemical Engineers, Warwickshire, UK.

Chapter Eight

Site restoration techniques

There are several proven remediation technologies and processes that may be employed in the management of contaminated site problems. The restoration methods often employed in contaminated site remediation programs may be classified as variations or combinations of physical, chemical, biological, and/or thermal treatment techniques. Most types of contaminant treatment processes and restoration technologies commonly used to remediate contaminated sites usually will belong to one or more of the following categories and examples:

- Biological treatment
  - Bioremediation (*in situ* or *ex situ*)
  - Composting
  - Landfarming
  - Phytoextraction
- Chemical treatment
  - Hydrogen peroxide catalysis
  - *In situ* chemical treatment (various, including precipitation, neutralization, oxidation-reduction [redox], and ion exchange)
- Physical treatment
  - Activated carbon adsorption
  - Soil flushing or washing
  - Vacuum extraction
  - Solvent extraction
  - Air or steam stripping
  - Electrodialysis
  - Reverse osmosis
- High-temperature treatment
  - Incineration
  - Infrared thermal treatment
  - *In situ* vitrification
  - Plasma-fired reactor
  - Electric reactor
- Low-temperature treatment
  - Low volatilization
  - *In situ* radio frequency heating
  - Low temperature thermal desorption
- Electrical treatment
  - Electrokinetic techniques
- Isolation and control by geotechnical methods
  - Hydraulic isolation (by pumpage and gradient control)

- Physical isolation by installation of barriers (e.g. capping and slurry walls)
- Physical isolation by hydraulic control measures (e.g. trenches and wells)
- Excavation and containment (e.g. landfilling)
- Reactive barriers (that create conditions for contaminant attenuation in a permeable barrier).
- Immobilization techniques
  - Encapsulation
  - Stabilization/solidification
  - Asphalt incorporation
- Material re-use without treatment
  - Recycling of contaminated materials.

A variety of remediation techniques are indeed available for contaminated site restoration programs. However, no one particular technology or process is usually appropriate for all contaminant types and/or under the variety of site-specific conditions that exist at different project sites. The choice of a remediation strategy and technology is driven by site conditions, contaminant types, sources of contamination, source control measures, and potential impacts of the possible remedial alternatives.

Table 8.1 provides a summary description of selected site restoration techniques that may be applied to the major environmental matrices contaminated by a variety of chemical compounds. In general, no single technology is universally applicable with equal success to all contaminant types and at all sites. Oftentimes, more than one remediation technique is needed to effectively address most contaminated site problems. In fact, treatment processes can be, and usually are, combined into process trains for more effective removal of contaminants and hazardous materials present at contaminated sites. For example, whereas biological treatment (with or without enhancement techniques) could result in the most desirable treatment scenario for petroleum-contaminated sites, the 'hot spots' at such sites may best be handled by physical removal (i.e. excavation) and thermal treatment of the removed materials – rather than by biological methods. Consequently, several technologies (or combination of technologies) that can provide both efficient and cost-effective remediation should normally be reviewed and explored as possible candidates in a remedy selection process.

This chapter offers a broad overview of several site restoration techniques that may be employed in a variety of contaminated site cleanup programs, recognizing that no one specific technique may be considered as a panacea to all contaminated site problems. A more detailed description of the processes, equipment and controls, as well as the detailed design elements of the various technologies, can be found elsewhere in the literature (e.g. Alexander, 1994; Anderson, 1993–95; API, 1993; Avogadro and Ragaini, 1994; ARB, 1991; Cairney, 1987, 1993; Charbeneau et al., 1992; Demars et al., 1995; Hinchee et al., 1994; Jolley and Wang, 1993; NRC, 1994; Nyer, 1992, 1993; OBG, 1988; Pratt, 1993; Rumer and Ryan, 1995; Schepart, 1995; Sims, 1990; Sims et al., 1986; USEPA, 1984, 1985, 1988, 1989, 1990; Vandegrift et al., 1992).

8.1 ACTIVATED CARBON ADSORPTION

The *activated carbon adsorption* technology is based on the principle that certain organic constituents preferentially adsorb to organic carbon. In the process involved,

Table 8.1 A summary listing of selected site restoration techniques

Contaminated environmental matrix	Remediation option	Basic technology and process	Scope of potential applications	Limitations/comments
Soils	Asphalt batching	Asphalt incorporation, involving the incorporation of petroleum-contaminated soils into hot asphalt mixes as a partial substitute for stone aggregate	• Economical for larger volumes of contaminated materials • Dependent on climate; most asphalt plants do not operate during cold weather	• May be unsuitable for clays • May require offsite transportation • May require specific analyses to be accepted for incorporation
	Bioremediation/ biorestoration	Natural or enhanced biodegradation, involving a process to degrade organic compounds into innocuous materials. Needs shallow groundwater (<15 meters [or 50 feet]) or an underlying impermeable silt or clay layer	• Most cost-effective for large volumes of contaminated material • Most applicable when contamination extends to groundwater, and is of sufficient volume or depth below surface • Minimal disruption to site operations, for *in situ* bioremediation	• Labor-intensive; requires considerable maintenance • Possibility for contaminant migration • Loss of efficiency in soils containing certain chemicals and low pH • Not effective in soils with low hydraulic conductivity, for *in-situ* bioremediation
	Encapsulation	*In situ* containment and isolation process that comprises of isolating contaminated soils from the surrounding environment by use of clay caps, liners, slurry walls, grout curtains, etc.	• Used only to prevent contaminant migration • Applicable to most chemicals, provided such compound does not attack containment materials • Open areas preferred • Example application involves capping landfills to prevent leaching by recharge water	• Long-term monitoring required • Does not destroy contaminants; only prevents migration • Solution may not be permanent
	Chemical fixation	Solidification/stabilization treatment process, involving the addition of materials to decrease the mobility of the original waste constituents	• Applicable to a wide variety of waste materials	• Long-term monitoring required • Future land-use may be restricted

Method	Description	Advantages	Disadvantages
Excavation/landfill disposal	Process involves the removal of contaminated soils	• Best for removal and disposal of limited volumes of 'hot spot' materials from a vast area that may otherwise be 'clean' • Common practice for shallow, highly-contaminated soils	• Potential for long-term liabilities • Long-term monitoring required • Trend is towards increasing disposal costs
Incineration	Thermal treatment	• Economy of scale for large volumes • Applicable to a broad range of organic compounds	• Generally high costs • Potential disposal problems for residual materials; residual ash may require further treatment
In situ soil leaching/flushing	In situ soil leaching process that involves injecting or flushing in-place soils with water to leach chemicals in soils into groundwater; surfactants may be added to facilitate the flushing process	• Applicable to both organics and inorganics, but to different degrees • Most applicable when contamination extends to groundwater table, requiring use in conjunction with extraction and treatment systems • Site hydrogeology has strong influence	• Leachate collection required • Long-term monitoring required • Potential problems with leaching fluid used • Less feasible for complex mixtures of several waste types • Costs depend on site characteristics, contaminant constituents, and cleanup levels • Not effective in soils with low hydraulic conductivity
Landfarming	Land treatment process, by which affected soils are removed and spread over a treatment area in a layer so as to enhance naturally-occurring degradational processes	• Best suited for lighter organic compounds • Warmer temperatures are conducive to faster degradation rates, since temperature influences the rate of degradation	• Emissions control is difficult • Low cleanup levels may not be practical • Requires relatively large treatment areas • Certain chemicals may be toxic to native microbes that could be facilitating degradation process

(continued)

Table 8.1 (*continued*)

Contaminated environmental matrix	Remediation option	Basic technology and process	Scope of potential applications	Limitations/comments
	Passive remediation	A 'no action' option that relies on several natural processes to destroy the contaminants of concern	• Site-specific: greater depth to groundwater, presence of aquitard, and low infiltration may minimize migration to groundwater • Temperature may affect volatilization and natural degradation	• Long-term monitoring required • Long-term liabilities possible • Low restoration levels may not be possible • Effectiveness may be influenced by soil conditions • Possible future land-use restrictions
	Vapor extraction systems	*In situ* volatilization, involving the removal of volatile organic contaminants from subsurface soils by mechanically drawing air through the soil matrix	• Applicable to volatile organic compounds (VOCs) • Can be applied to wide areas • Minimal disruption to site operations; conducive to both developed and undeveloped sites • Most effective at higher concentrations • Generally low costs	• Venting and emissions difficult to control for very shallow areas • Not effective below the groundwater table • Performance affected by soil conditions; less effective and/or longer time frame in fine-grained soils • May not work for semi-VOCs
Groundwater	Air stripping	Aeration process, effective for removal of volatile organic contaminants	• Simple technology, usable in conjunction with other methods • Low maintenance costs	• Problems with air emissions; collection and treatment systems required for offgas releases

Method	Description	Advantages	Disadvantages
Carbon adsorption	Activated carbon adsorption technology, based on the principle that certain organic contaminants preferentially adsorb on to organic carbon	• Simple technology, usable in conjunction with other methods • Attractive as point-of-use treatment method • More appropriate for aquifer restoration • Applicable to a broad range of (organic and inorganic) contaminants	• Generates 'spent carbon' as a waste byproduct • High maintenance costs; carbon requires frequent regeneration
Groundwater extraction and treatment	Groundwater pump-and-treat system for the restoration of contaminated aquifers; it involves pumping groundwater for treatment at the surface	• Very common approach for the restoration of contaminated groundwater • Best used for containing contaminant plumes	• Treated water to be discharged or re-injected may require compliance with certain pollutant discharge standards or criteria • Generally requires use of air pollution control devices to remove contaminants released into air • Not considered cost-effective nor efficient for aquifer restoration
Passive remediation	A 'no action' option that relies entirely on several natural processes to destroy the contaminants of concern	• Several physical and chemical variables (e.g. temperature and alkalinity) may affect natural degradation and other attenuation processes	• Long-term monitoring required • Long-term liabilities possible • Low restoration levels may not be possible

granular activated carbon (GAC) is packed in vertical columns, and contaminated water flows through it by gravity. GAC has a high surface area to volume ratio, and many compounds readily bond to the carbon surfaces. Contaminants from water are therefore adsorbed to the carbon, and the effluent water has a lower contaminant concentration; spent carbon (i.e. carbon that has reached its maximum adsorption capacity) from the GAC adsorption is typically regenerated by incineration. Water may be passed through several of these columns to complete contaminant removal.

Activated carbon systems are generally capable of efficiently removing very low concentrations of dissolved organics from groundwater.

8.2   ASPHALT BATCHING

Asphalt batching (or *asphalt incorporation*) is a method for treating hydrocarbon-contaminated soils. It involves the incorporation of petroleum-laden soils into hot asphalt mixes as a partial substitute for stone aggregate. This mixture can then be utilized for pavings.

The process of asphalt batching consists of excavating the contaminated soils which then undergo an initial thermal treatment, followed by incorporation of the treated soil into aggregate for asphalt. During the incorporation process, the mixture (including the impacted soils) is heated, resulting in the volatilization of the more volatile hydrocarbon constituents at various temperatures. The remainder of the compounds become incorporated into the asphalt matrix during cooling, thereby limiting constituent migration.

8.3   BIOREMEDIATION

Bioremediation (or *biorestoration*) has become a viable and cost-effective site restoration technology for treating a wide variety of contaminants (such as petroleum and aromatic hydrocarbons, chlorinated solvents, and pesticides). As an innovative technology, bioremediation offers the very important advantage of permanently removing contaminants by biodegradation, and therefore reduces potential long-term liabilities (since the contamination is destroyed rather than removed in the biorestoration process).

Biorestoration relies on microorganisms (especially bacteria and fungi) to transform hazardous compounds found in soil and groundwater systems (or even air streams) into innocuous or less toxic metabolic products. An optimized biotransformation condition may be attained by manipulating the physical environment and controlling nutrient supplements. The technique requires careful process control to establish the appropriate microbial population. By using microorganisms (natural or engineered) to degrade contaminants in soil, groundwater, or air, bioremediation transforms hazardous/toxic materials into non-toxic elements such as water, carbon dioxide, and other innocuous products.

In general, the biodegradation of a compound under field conditions is affected by temperature, type of target compounds, soil dissolved oxygen levels, soil moisture content, soil permeability, oxidation–reduction potential, soil pH, compound availability and concentrations, availability of nutrients, and the natural microbial community. These factors act together to determine the biodegradability of the contaminants in a particular setting.

Since most contaminated sites require prompt remedial actions, an acceleration of the natural biodegradation process (i.e. enhanced biodegradation) is generally desirable. Under such circumstances, the soil system may have to be modified to promote the activity of the naturally found organisms. Promotion methods may include the addition of nutrients and aeration of the soil. If the intrinsic microorganism flora do not work on specific types of contaminants, selectively adapted inoculants can also be added to the soil. The enhanced biodegradation process may be achieved as a conventional waterborne biodegradation, in which water is used to uniformly transport nutrients (nitrogen and phosphorus) and oxygen through the unsaturated zone; or as a soil venting enhanced biodegradation, in which a system is engineered to increase the microbial biodegradation in the vadose zone utilizing pumped air as the oxygen source; or as a variation of a conventional waterborne and/or a soil venting enhanced biodegradation.

8.4 CHEMICAL FIXATION/IMMOBILIZATION

Chemical fixation immobilization (or *stabilization*) is a technique to chemically fix or modify the chemical structure of contaminants, by applying specific reagents. The fixation process consists of immobilizing the contaminants in-place, thereby preventing migration into unaffected environmental media. Chemical fixation can be applied to many organic and metal bearing contaminants or wastestreams. In its application, it generally is necessary to perform bench-scale treatability studies, prior to a field/full-scale treatment.

In the chemical fixation technology, the contaminated soil or material is blended with precise amounts of reagent(s) to stabilize and/or encapsulate chemical constituents which are then stockpiled and allowed to cure. Oftentimes, the treated material is rendered non-hazardous and may be backfilled and left on the site. The process generally improves soil condition – such as an increased compressive strength or decreased permeability. Thus the stabilized soil may also be used as structural fill material in geotechnical construction activities, without posing significant threats to the surrounding environment.

8.5 ENCAPSULATION (BY PHYSICAL BARRIER SYSTEMS)

Encapsulation (using physical barrier systems) is a remedial alternative comprised of the physical isolation and containment of the contaminated material, as is typical of a well-engineered landfill. The encapsulation technique consists of isolating the impacted soils through the use of low permeability caps, slurry walls, grout curtains, or cut-off walls. The contaminant source is covered with low-permeability layers of synthetic textiles or clay cap; the cap is designed to limit infiltration of precipitation and thus prevent leaching and migration of contaminants away from the site and into groundwater.

The encapsulation technique often includes a physical barrier installation at the sides of the contaminated area because its purpose is to leave the contaminant safely in-place at the location of choice. In general, isolation and containment systems can work adequately, but there is no guarantee as to the destruction of the encapsulated contaminants.

8.6 *IN SITU* LEACHING

In situ soil leaching (or *in situ flushing*) is a process by which in-place soils are flushed with water, usually mixed with a biodegradable nontoxic surfactant, in an effort to leach the compounds present in the soil into groundwater. The flushing agent is allowed to percolate into the soil and enhance the transport of contaminants to groundwater extraction wells for recovery. The groundwater is then collected downgradient or downstream of the leaching site for treatment, recycling, and/or disposal.

In a typical application, inorganic contaminants can be extracted from soil by flushing the soil with solvents; solvents are recovered, contaminants are extracted, and the solvents are re-circulated through the soils. In general, a high water solubility, a low soil–water partition coefficient (K_{oc}), and a porous soil matrix will aid in the effective removal of chemical contaminants from soils using the soil leaching technique. The K_{oc}, which is a measure of the equilibrium between the soil organic content and water, is indeed the leading factor controlling the effectiveness of soil flushing; a low K_{oc} value indicates a favorable leaching tendency of the constituent from the soil.

Water is normally used as the flushing agent in the *in situ* leaching technique. However, other solvents may be used for contaminants that are tightly held or only slightly soluble in water. Chemically enhanced *in situ* soil flushing may have extensive applications, but such applications will generally require site-specific evaluation and a relatively more sophisticated system design.

For hydrophobic compounds, flushing with surfactants is likely to be more effective than flushing with water, whereas flushing with water generally may suffice for hydrophilic compounds. When used, solvents are selected on the basis of their ability to solubilize the contaminants and also on their environmental and human health effects. Thus, it is important to know the chemistry and toxicity of the surfactant. It is also important to understand the hydrogeology of the site to ensure that contaminants will be extracted once they are mobilized.

8.7 SOIL WASHING

Soil washing (or *solvent washing* or *solvent/chemical extraction*) consists of excavating soils from contaminated areas and washing the contaminants from the soil using water or an aqueous solution. The soil washing technique is most applicable for soluble organic chemicals and metals. Example contaminant groups that may be effectively dealt with by soil washing include petroleum and fuel residues, heavy metals, PCBs, PAHs, pesticides, creosote, and cyanides (Pratt, 1993). Typically, inorganic and organic contaminants are extracted from the soil by washing the soil with solvents; the contaminated effluent is then recovered, contaminants are extracted, and the solvents are re-circulated through the soils or disposed of.

Water is normally used as the washing agent; however, other solvents may be used for contaminants that are tightly held or only slightly soluble in water. Solvents are selected on the basis of their ability to solubilize the contaminants and also on their environmental and human health effects. Thus, it is important to know the chemistry and toxicity of the surfactant of choice.

8.8   THERMAL TREATMENT TECHNOLOGIES

Thermal treatment technologies basically employ heat to destroy or change the contaminants of concern present in a contaminated material. Some of the more common thermal treatment alternatives finding extensive application in site restoration programs are annotated below.

8.8.1   Incineration

Incineration is a thermal treatment/degradation process by which contaminated materials are exposed to excessive heat; it typically involves the thermal destruction of contaminants by burning. Of particular interest is *catalytic conversion*, which is an incineration process that uses a catalyst to reduce the usually high incineration temperature requirements. Used in special treatment applications, catalytic units can be built for specialized contaminants, but the cost of such units tends to be rather high.

Depending on the intensity of the heat, the contaminants of concern are volatilized and/or destroyed during the incineration process. In fact, incineration of high calorific value wastes may be regarded as a form of recycling if the heat generated is utilized for other economic purposes. The ash remaining from incineration of hazardous materials is usually disposed of in landfills.

8.8.2   *In Situ* Vitrification

In situ vitrification (ISV) is an innovative thermal treatment process that converts contaminated materials into a chemically inert, stable glass and crystalline product. The ISV method may be used to provide solution to mixed wastes (including organic, inorganic, and radioactive wastes) in soils at a potentially contaminated site, up to about a 15-meter (≈ 50-foot) depth. During the vitrification process, the major portion of the contaminants initially present in the soils is volatilized, with the remainder being worked in-place in the hardened soil. The ISV process uses electrical energy to melt contaminated solids at very high temperatures.

The initial step in the vitrification process is to identify the boundaries of the area of contaminated soil to be treated. An array of electrodes is inserted into the ground to the desired treatment depth. The electrodes are placed vertically into the contaminated soil region, and an electrical potential of over 12 kilovolts is applied to the electrodes which establishes an electrical current. The resultant power heats the path and surrounding soil to above fusion temperatures. The soil is melted by the resulting high temperatures. When the melt cools and solidifies, the resulting material is stable and glass-like, with the contaminants bound in the solid.

8.8.3   Thermal Desorption

Thermal desorption is the separation of contaminants from a solid matrix through volatilization processes. It consists of an *ex situ* means for physically separating organic compounds from soils, sediments, sludges, filter cakes, and other solid media (Anderson, 1993). The process does not normally remove or stabilize metals that may be present in the feed material. However, since the units operate at relatively low temperatures, they do not oxidize the metals either. Thus, in general, the metals do *not* become more mobile or leachable after treatment by thermal desorption – as opposed to what might happen in the case of incineration.

Thermal desorbers are not necessarily designed to effect decomposition; depending on the nature of the organic chemicals present and the temperature of the system, however, some decomposition may occur. In any case, the desorber performance is generally measured by the difference between contaminant levels in the untreated medium *versus* the contamination levels remaining in the 'processed' medium.

Thermal desorption is often used as part of a treatment train, which means that some pre- and/or post-processing will usually be necessary in its application. In the processes involved, the contaminated material is excavated and delivered to the thermal desorber. Treated material may be suitable for return to the site of origination.

8.9   ULTRAVIOLET-OXIDATION TREATMENT TECHNOLOGIES

Ultraviolet (UV)-oxidation treatment methods represent one of the most important technologies emerging as viable treatment techniques for groundwater remediation (Nyer, 1993). The systems involved generally use an oxygen-based oxidant (e.g. ozone or hydrogen peroxide), in conjunction with UV light. The process involves placing UV bulbs in a reactor where the oxidant comes into contact with the contaminants in the groundwater, but the bulbs cannot come into direct contact with the water.

Oxidation technologies can indeed be used to completely destroy many organic compounds that may be present in contaminated waters. UV-oxidation is mostly applied in the following two basic forms:

- *UV-Peroxide Systems* (i.e. oxidation with ultraviolet light and hydrogen peroxide), in which high-intensity ultraviolet lights are used to catalyze the formation of hydroxyl radicals from hydrogen peroxide. Under controlled conditions, the hydroxyl radical (which is a very powerful chemical oxidant) reacts with the contaminants present, oxidizing the chemicals into less harmful compounds.
- *UV-Ozone Systems* (i.e. oxidation with ultraviolet light and ozone), which utilizes the strong oxidizing properties of UV light and ozone – a combination with superior destructive results, in comparison to the use of either oxidant alone. The UV light and ozone act synergistically to oxidize the contaminants present.

It is noteworthy that, whereas ozone and hydrogen peroxide are both strong oxidizing agents by themselves, their effectiveness increases dramatically when stimulated by UV light (Nyer, 1993). In general, there are no air emissions or waste by-products from the process if the reactions are carried to completion.

8.10 VAPOR EXTRACTION SYSTEMS

Vapor extraction systems (VESs) may be applied to the removal of volatile organic chemicals (VOCs) present at contaminated sites. The VES technology is a particularly economical and efficient means of removing VOCs from the subsurface environment. In fact, it may even be suitable for removing the volatile components of NAPLs in the unsaturated zone that are trapped at residual saturation, and also free product layers. In any case, the VES is most applicable to the remediation of the higher volatile or lower molecular weight constituents of organic compounds only. As a general rule, heavier organic fractions (such as diesel fuel and fuel oils) are not candidates for remediation by vapor extraction. Also, vapor extraction is more effective at sites where the more volatile chemicals are still present, especially when the release is reasonably recent.

In general, the VES may vary considerably in size and design, depending on site-specific requirements and/or conditions. Typically, a well-designed VES consists of a series of extraction/injection wells connected to a common manifold and a positive displacement air blower, together with other surface equipment (such as emission controls, instrumentation, electric motor, etc.). A very important advantage for using a VES relates to the fact that there is minimal site disruption during its implementation/operation. VESs can indeed be designed for many areas of site remediation such as contaminated soil piles and inaccessible locations (e.g. underneath buildings and similar structures). The most common types of VES applications include soil vapor extraction, air or steam stripping, and air sparging; these systems are discussed below.

8.10.1 Soil Vapor Extraction

Soil vapor extraction (SVE) (or *in situ soil venting, vacuum extraction,* or *in situ soil stripping*) is a technique that uses soil aeration to treat subsurface zones of VOC contamination in soils. In the SVE technology, VOCs are extracted from soil by using a vacuum system. This soil cleanup technique employs vacuum blowers to pull large volumes of air through contaminated soil. The air flow flushes out the vapor-phase VOCs from the soil pore spaces, disrupting the equilibrium that exists between the contaminants on the soil and in the vapor. This causes further volatilization of the contaminants and subsequent removal in the air stream. Treatment rate depends on air flow through the soils and how effectively the contaminants partition into the mobile air phase (Newman et al., 1993).

The SVE process involves removing and venting VOCs from the vadose or unsaturated zone of contaminated soils by mechanically drawing or venting air through the soil matrix. Fresh air is injected or allowed to flow into the subsurface at locations in and around the contaminated soil to enhance the extraction process. This is carried out by connecting a vacuum pump or fan to one or more extraction wells, with the extraction wells typically being installed to penetrate the contaminant plume near the zone of highest VOC concentration. When suction is applied to the extraction wells, it induces a subsurface radial air flow towards perforations in the well casings. In addition, injection or ventilation wells (to facilitate the infiltration of clean air into the soil) may be placed at selected locations to help direct the flow of induced air towards the extraction wells. The inlet air going to the injection well may be supplied by the

VOC control treatment exhaust, or the vacuum pump (blower) exhaust at the site engineer's discretion. The VOC-laden air is withdrawn under vacuum from recovery or extraction wells which are placed in selected locations within the contaminated site. This air is then either vented directly to the atmosphere, or it is vented to an above-ground level VOC treatment unit such as a carbon adsorber or a catalytic incinerator prior to being released to the atmosphere. The decision to employ VOC control system treatment is largely dependent on VOC concentrations and applicable local environmental regulations. The selection of a particular VOC treatment option may be based in part on individual site characteristics.

SVE is generally viewed as a cost-effective technique for VOC removal, and is therefore finding more widespread application. The method is a relatively simple concept and can be used in conjunction with other soil decontamination procedures such as biological degradation. It should be noted, however, that because contaminants must volatilize and partition into air undergoing removal, contaminated sites having significant amounts of low-volatility compounds (e.g. diesel fuel, fuel oils, or jet fuels) often are not targeted for remediation by soil venting. Sites having soil heterogeneities that result in uneven air permeation also can prevent effective remediation by conventional soil venting.

Enhancing the performance of SVE systems   Removal of VOCs from vadose zone soils through vacuum extraction has become an important remedial alternative of choice for many contaminated sites. The removal efficiency at contaminated sites can generally be enhanced in a variety of ways, including the following:

- By use of temporary or permanent caps over the contaminated soil; if low permeability strata are present or if other, non-volatile contaminants are encountered with the VOCs, a permanent cap may be more cost-effective. Capping the entire site with plastic sheeting, clay, concrete or asphalt enhances horizontal movements towards the extraction vent. Impermeable caps extend the radius of influence around the extraction vent. The use of a ground surface cover will also prevent or minimize infiltration, which, in turn, reduces moisture content and further chemical migration.
- Overall costs of SVEs can be reduced and cleanup accelerated by applying heat to enhance vaporization. When a heat pump arrangement is installed, product recovery can be achieved along with better granular activated carbon (GAC) adsorber performance (where GAC is employed) and reduced operating costs. Furthermore, the restoration time-frame can be reduced if vaporization and diffusion rates are increased by heating. An additional advantage of heating is the possibility or ability to remove less volatile chemicals.

SVE is indeed effective under a wide range of site conditions. Insulation is occasionally used on the piping and headers, especially in colder climates, to prevent condensate freezing. Extraction vents/wells are typically designed to fully penetrate the unsaturated soil zone or the geologic stratum to be cleaned. Spacing of extraction vents is usually based on an estimate of the radius of influence of an individual extraction vent; vent spacing typically ranges from about 4.5 to 30 meters (≈ 15 to 100 feet) (Hutzler et al., 1990). If water should be pulled from the extraction vents, an air–water separator is required to protect the blowers or pumps and to increase the efficiency of vapor treatment systems. In general, SVE systems offer a relatively inexpensive means of removing chemicals without major disruption to the project site.

8.10.2 Air Stripping

Air stripping is a remediation technique that involves the physical removal of dissolved-phase contamination from a water stream. It is a separation technology that takes advantage of the fact that certain chemicals are more soluble in air than in water (Nyer, 1993). By bringing large volumes of air in contact with the contaminated water, therefore, a driving gradient from the water to the air can be created for the contaminants of concern.

The air stripping process basically involves pumping contaminated water containing VOCs from the ground and allowing it to trickle over packing material in an air stripping tower. At the same time, clean air is circulated past the packing material. When the contaminated water comes into contact with the clean air, the contaminants tend to volatilize from the water into the air. The contaminated air is then released into the atmosphere, or a GAC system. Naturally, the removal efficiency of an air stripper depends upon the volatility of the contaminants.

Air strippers are designed to maximize removal of VOCs from groundwater, leading to transfer of these contaminants into air that can be treated in a control device or discharged to the atmosphere. In a typical air stripping system, contaminated water containing VOCs is counter-currently contacted with air in a packed tower. By contacting contaminated water with clean air, dissolved VOCs are transferred to the airstream to create equilibrium between the phases. The process takes place in a cylindrical tower, packed with inert material which allows sufficient air/water contact to remove volatiles from water. Contaminants are then removed from the airstream. An optional unit operation (e.g. catalytic oxidation or vapor-phase carbon adsorption) may be used for the control of air emissions. The stripped water may further be treated in an optional carbon absorber polishing bed. The treated effluent water is either recycled as process water or discharged; in the case of a groundwater cleanup operation for site restoration programs, the treated water may be pumped back into the aquifer.

Air stripping of chemical contaminants from contaminated water is an effective method of removing VOCs from the contaminated water. However, this method also transfers pollutants from the water to the gas phase, and the resulting air emissions may need to be appropriately controlled. Indeed, air stripping is not a destruction technology; it merely moves contaminants from a liquid phase to an air phase, to be addressed differently.

8.10.3 Air Sparging

Air sparging is the highly controlled injection of air into a contaminant plume in the soil *saturated* zone. Air pumped into contaminated groundwater is used to strip volatiles from the groundwater to the soil vadose zone for capture using SVE. Air bubbles traverse horizontally and vertically through the soil column, creating a transient air-filled porosity in which volatilization can occur. In fact, air sparging effectively creates a crude air stripper in the subsurface, with the soil acting as the packing.

In a typical design of an air sparging system, an array of vents (or shallow wells) penetrating the impacted area of the vadose zone is connected via manifolding to air blowers. The blowers create a partial vacuum in the vents and pull air, including VOCs, out of the soil. The air sparging treatment involves injecting the soils with air, which flows vertically and horizontally to form an oxygen-rich zone in which VOCs are volatilized. Air bubbles that contact dissolved or adsorbed-phase contaminants in the aquifer cause the VOCs to volatilize. The volatilized organics are carried by the air bubbles into the vadose zone where they can be captured by a VES. Also, the sparged air maintains a high dissolved oxygen (DO) content, which enhances natural biodegradation. A carbon treatment system is often used to treat off-gases. Careful investigation for optimal system design will result in an efficient operating system.

Used in conjunction with soil vapor extraction, air sparging is emerging as an effective treatment technology for soils and groundwater contaminated with VOCs (see Section 8.13). Despite the fact that air sparging systems generally require very sophisticated analysis of site hydrogeology and careful engineering before implementation, overall, the system can be very cost-effective.

8.11 BIOVENTING

Bioventing is a variation of the VES. It comprises of the delivery of oxygen to unsaturated soils by forced air movement for the purpose of enhancing biodegradation of organic contaminants. In bioventing, increased microbial activity results in the degradation of contaminants that are less easily removed by volatilization using the VES.

The bioventing process injects air into the contaminated media at a rate designed to maximize *in situ* biodegradation and minimize or eliminate off-gassing of volatilized contaminants to the atmosphere. In fact, this modification of the conventional vapor extraction remediation process to allow contaminants to be biologically removed *in situ* will reduce or eliminate air emissions, and therefore significantly cut down remediation costs. Bioventing also biodegrades less volatile organic contaminants and allows treatment of less permeable soils, because a reduced volume of air is required for treatment (Newman et al., 1993).

8.12 GROUNDWATER PUMP-AND-TREAT SYSTEMS

By and large, the most common method of groundwater restoration has been the application of *pump-and-treat systems*. The performance of these systems depends directly on site conditions and contaminant chemistry (NRC, 1994).

Groundwater pump-and-treat systems involve contaminated groundwater being pumped out of the ground, treated by an appropriate treatment method to remove the contaminants of concern, and re-injected into the ground or used otherwise. Technologies for groundwater extraction and treatment are generally used to address the treatment needs of site-specific conditions and regulatory requirements. A common approach is to combine technologies to achieve effective treatment and to meet discharge criteria.

Invariably, groundwater extraction and treatment as a remedial action must address issues pertaining to the strategic and optimum design of the extraction–injection well network, as well as the selection of the proper treatment technology for the extracted groundwater. In general, pump-and-treat may be more appropriate for containing contaminant plumes, or for use in initial emergency response actions at sites having nonaqueous phase liquid releases to groundwater. If free product hydrocarbons are present, then an oil/water separator may be required as part of the overall restoration program.

Groundwater pumping actions can have a number of configurations and design objectives. Single pumping wells or a line of well points can be used to capture a plume. Single or multiple wells can be installed to divert groundwater by lowering the water table; they can also be used to prevent unconfined aquifers from contaminating lower aquifers separated by leaky formations. The water withdrawn by pumping may be treated and subsequently re-injected through one or more wells. The re-injection wells may be used to flush contaminants toward the pumping wells or to create a hydraulic barrier to preclude further plume migration.

8.13 COMBINED AIR SPARGING AND SVE SYSTEMS

A variety of *in situ* techniques have been used in attempts to remediate VOC-contaminated sites. However, most of the techniques have proved to be of limited effectiveness when used to remediate saturated soils and groundwater. The use of an *in situ* air sparging system in conjunction with a SVE system is becoming increasingly popular as a more efficient and cost-effective method for the remediation of VOC-contaminated saturated soils and groundwater (Reddy et al., 1995). These systems (consisting of air injection wells, vapor extraction wells, and several other appurtenances) are generally most effective for localized contamination of known extent. Air injected into the saturated zone transports the contaminants to the top of the saturated zone and into the unsaturated zone, to be captured through the SVE extraction wells by an induced vacuum. When designed and operated properly, this combination system can prove very cost-effective in the restoration of contaminated saturated soils and groundwater.

In fact, air sparging, in combination with SVE, has been used to remove and destroy organic solvents at some difficult corrective action cleanup sites. A classical example involves the cleanup of VOCs adsorbed on to soils below the water table at a facility located adjacent to a wetlands area. Under such circumstances, applying SVE alone would require pumping out groundwater to lower the water table; however, this option will be unacceptable because of the presence of the wetlands. Consequently, an air sparging–soil vapor extraction design may be used in these types of situations. With air sparging, air injected into saturated soils travels vertically and horizontally to form an oxygen-rich zone in which adsorbed and dissolved VOCs are volatilized. As vapors rise from the saturated zone to the soil vadose zone above, VOCs are captured by the SVE system.

8.14 ELECTROKINETIC SOIL DECONTAMINATION

The use of electrokinetic soil decontamination techniques represents recent advances made in several fronts in the arena of contaminated site restoration. Electrokinetic soil decontamination uses electrodes to pass a direct current through a contaminated soil, causing ions and water to migrate through the subsurface toward the electrodes (Acar and Alshawabkeh, 1993; Acar et al., 1993; Alshawabkeh and Acar, 1992). Small pumps placed at the cathodes remove the metal-containing fluids for treatment. Electrokinetics generally is applied only to relatively impermeable soils (e.g. fine-grained soils).

8.15   PHYTOREMEDIATION/PHYTOEXTRACTION

Phytoremediation (or, *phytoextraction*) consists of the use of vegetation for the *in situ* treatment of contaminated soil and sediments. This emerging technology is most suited for sites with shallow contamination (usually <5 meters [≈17 feet] depth), and it includes the use of plants to remediate sites contaminated by metals (and in some cases organic compounds) (Kumar et al., 1995; Schnoor et al., 1995). Reasonable advances have indeed been made in the use of phytoremediation to address contaminated site problems.

Phytoremediation uses plants known as hyperaccumulators and halophytes to remove metals from a contaminated soil. These plants show preferential uptake of metals, accumulating them in their tissues. Typically, the high metal accumulation by some plants suggests that such plants may be used to clean up toxic metal-contaminated sites in this process of phytoextraction. In fact, a small number of wild plants that grow on metal-contaminated soils accumulate large amounts of heavy metals in their roots and shoots; some of these species can accumulate unusually high concentrations of toxic metals to levels which far exceed the soil levels. This property may therefore be exploited for soil reclamation if an easily cultivated, high biomass crop plant able to accumulate heavy metals is identified (Kumar et al., 1995).

The process of phytoextraction generally requires the translocation of heavy metals to the easily harvestable shoots, and in some cases the roots as well. Subsequently, dried, ashed, or composted plant residues highly enriched in heavy metals may be isolated as hazardous waste or recycled as metal ore (Kumar et al., 1995). It is noteworthy that, whereas very heavily contaminated soils may not support plant growth, sites with light to moderate toxic metal contamination could be remediated by growing metal-accumulating plants. Each cleanup situation may indeed require a different plant species or a number of plants in tandem. Generally, phytoremediation is used in conjunction with other cleanup approaches in the site restoration effort.

8.16   FREE PRODUCT RECOVERY

The recovery of free product hydrocarbon or solvents floating atop a groundwater table (i.e LNAPL) is similar in concept to the pump-and-treat system. For example, LNAPLs can be removed using physical recovery techniques such as a single pump system that produces water and free product, or a two-pump/two-well system that steepens the hydraulic gradient and recovers the accumulating free product.

On the other hand, special circumstance problems call for greater innovation in methods of choice. For example, there generally are no good methods for the recovery of DNAPLs. Remedial strategies for DNAPL sites typically involve containment, removal, or a combination of both. Whenever DNAPL pools are present, environmental regulations almost invariably mandate the removal of the free product (usually by product pumpage). Removing DNAPL residual is more difficult, especially when it is below the water table. The diagnosis and assessment of DNAPL sites indeed presents itself as a very complex issue, making remediation decisions at such sites even more challenging – as recognized by the limited literature on this subject matter (e.g. Pankow and Cherry, 1996).

8.17 PASSIVE REMEDIATION

Passive remediation relies on natural processes (e.g. biodegradation, volatilization, photolysis, sorption, dispersion and dilution) to remediate impacted soils and groundwater. Passive remediation may be applicable at sites where contaminant migration is limited, where potential impacts on the environment are minimal, and when health and safety considerations are insignificant.

When a passive remediation alternative is employed at contaminated sites, continued monitoring is usually used to demonstrate that contamination levels are being attenuated, and that no significant exposure is occurring. Thus, following site assessment, the only activity undertaken is a progressive monitoring program to evaluate the effectiveness of the 'no-action' option in the management of a contaminated site problem.

8.18 REFERENCES

Acar, Y.B. and A. Alshawabkeh (1993). *Principles of Electrokinetic Remediation.* Journal of Environmental Science and Technology, **27**(13) (Dec. 1993), pp. 2638–2647.

Acar, Y.B., A.N. Alshawabkeh, and R.J. Gale (1993). *Fundamentals of Extracting Species from Soils by Electrokinetics.* Waste Management, Pergamon Press, London, **13**(2), pp. 141–151.

Alexander, M. (1994). *Biodegradation and Bioremediation.* Academic Press, San Diego, California.

Alshawabkeh, A. and Y.B. Acar (1992). *Removal of Contaminants from Soils by Electrokinetics: A Theoretical Treatise.* Journal of Environmental Science and Health, Part(a), **A27**(7), pp. 1835–1861.

Anderson, W.C. (ed.) (1993–95). *Innovative Site Remediation Technology.* Volumes 1–8. American Academy of Environmental Engineers, Annapolis, Maryland.

API (1993). *Guide for Assessing and Remediating Petroleum Hydrocarbons in Soils.* American Petroleum Institute, Washington, DC. API Publication No. 1629.

ARB (Air Resources Board) (1991). *Soil Decontamination.* Compliance Assistance Program, Air Resources Board, Compliance Division, California.

Avogadro, A. and R.C. Ragaini (eds) (1994). *Technologies for Environmental Cleanup: Toxic and Hazardous Waste Management.* Kluwer Academic Publishers, Dordrecht, The Netherlands.

Cairney, T. (ed.) (1987). *Reclaiming Contaminated Land.* Blackie Academic & Professional, Glasgow, UK.

Cairney, T. (ed.) (1993). *Contaminated Land (Problems and Solutions).* Blackie Academic & Professional, Glasgow/Chapman & Hall, London/Lewis Publishers, Boca Raton, Florida.

Charbeneau, R.J., P.B. Bedient, and R.C. Loehr (eds) (1992). *Groundwater Remediation.* Water Quality Management Library, Vol. 8. Technomic Publishing Co., Inc., Lancaster, Pennsylvania.

Demars, K.R., G.N. Richardson, R.N. Yong, and R.C. Chaney (eds) (1995). *Dredging, Remediation, and Containment of Contaminated Sediments.* ASTM Publication No. STP 1293, ASTM, Philadelphia, Pennsylvania.

Hinchee, R.E., B.C. Alleman, R.E. Hoeppel, and R.N. Miller (eds) (1994). *Hydrocarbon Bioremediation*. CRC Press/Lewis Publishers, Boca Raton, Florida.

Hutzler, N.J., J.S. Gierke, and B.E. Murphy (1990). Vaporizing VOCs. *ASCE Civ. Engnr.* **60**(4), April, 57–60.

Jolley, R.L. and R.G.M. Wang (eds) (1993). *Effective and Safe Waste Management: Interfacing Sciences and Engineering with Monitoring and Risk Analysis*. Lewis Publishers, Boca Raton, Florida.

Kumar, P.B.A.N. et al. (1995). Phytoextraction: the use of plants to remove heavy metals from soils. *Environ. Sci. Technol. ES&T* **29**(5), 1232–1238.

Newman, B., M. Martinson, G. Smith, and L. McCain (1993). 'Dig-and-Mix' bioventing enhances hydrocarbon degradation at service station site'. *Hazmat World*, December, 34–40.

NRC (National Research Council) (1994). *Alternatives for Ground Water Cleanup*. Committee on Ground Water Cleanup Alternatives. National Academy Press, Washington, DC.

Nyer, E.K. (1992). *Groundwater Treatment Technology*, 2nd edn. Van Nostrand Reinhold, New York.

Nyer, E.K. (1993). *Practical Techniques for Groundwater and Soil Remediation*. Lewis Publishers, Boca Raton, Florida.

OBG (O'Brien & Gere Engineers, Inc.) (1988). *Hazardous Waste Site Remediation: The Engineer's Perspective*. Van Nostrand Reinhold, New York.

Pankow, J.F. and J.A. Cherry (eds) (1996). *Dense Chlorinated Solvents and Other DNAPLs in Ground Water*. Waterloo Educational Services, Guelph, Ontario, Canada.

Pratt, M. (ed.) (1993). *Remedial Processes for Contaminated Land*. Institution of Chemical Engineers, Warwickshire, UK.

Reddy, K.R., S. Kosgi, and J. Zhou (1995). A review of in situ air sparging for the remediation of VOC-contaminated saturated soils and groundwater. *Hazard. Waste Hazard. Mat.* **12**(2), 97–112.

Rumer, R.R. and M.E. Ryan (eds) (1995). *Barrier Containment Technologies for Environmental Remediation Applications*. J. Wiley & Sons, New York.

Schepart, B.S. (ed.) (1995). *Bioremediation of Pollutants in Soil and Water*. ASTM Publication No. STP 1235, ASTM, Philadelphia, Pennsylvania.

Schnoor, J.L. et al. (1995). Phytoremediation of organic and nutrient contaminants. *Environ. Sci. Technol. (ES&T)* **29**(7), 318A–323A.

Sims, R.C. (1990). Soil remediation techniques at uncontrolled hazardous waste sites, a critical review. *J. Air Waste Mgmnt Assoc.* **40**(5), May, 704–732.

Sims, R. et al. (1986). *Contaminated Surface In-Place Treatment Techniques*. Noyes Publications, Park Ridge, New Jersey.

USEPA (US Environmental Protection Agency) (1984). *Review of In-place Treatment Techniques for Contaminated Surface Soils*, Vols 1 and 2, US Environmental Protection Agency, Hazardous Waste Engineering Research Laboratory, Cincinnati, Ohio. EPA-540/2-84-003a and b.

USEPA (US Environmental Protection Agency) (1985). *Modeling Remedial Actions at Uncontrolled Hazardous Waste Sites*. EPA/540/2-85/001 (April 1985). Office of Emergency and Remedial Response, Washington, DC.

USEPA (US Environmental Protection Agency) (1988). *Guidance on Remedial Actions for Contaminated Ground Water at Superfund Sites*. Office of Emergency and Remedial Response, Washington, DC. EPA/540/G-88/003.

USEPA (US Environmental Protection Agency) (1989). *Soil Vapor Extraction VOC Control Technology Assessment*. Office of Air Quality Planning and Standards, Research Triangle Park, North Carolina. EPA-450/4-89-017 (September 1989).

USEPA (US Environmental Protection Agency) (1990). *Air Stripper Design Manual*. Air/ Superfund National Technical Guidance Study Series. Office of Air Quality Planning and Standards, Research Triangle Park, NC. EPA-450/4-90-003 (May 1990).

Vandegrift, G.F., D.T. Reed, and I.R. Tasker (eds) (1992). *Environmental Remediation: Removing Organic and Metal Ion Pollutants*. ACS Symposium Series 509, American Chemical Society, Washington, DC.

Chapter Nine

Corrective measure evaluation tools

Corrective measure evaluation typically is comprised of a feasibility study of remedial options, the purpose of which is to examine site characteristics, cleanup goals, and the anticipated performance of alternative remedial technologies so that the most effective approach for the restoration of a contaminated site can be identified. A well-designed feasibility study addresses every contaminant migration pathway and environmental medium that poses, or could pose, unacceptable risks to human health or the environment. To accomplish the tasks involved in this type of evaluation, several corrective measure evaluation tools will usually be used to assist in determining whether or not remediation is necessary for a contaminated site problem, and to further determine the cleanup goals and techniques appropriate for a given site. This chapter highlights some representative analytical tools potentially applicable to the management of a variety of environmental contamination problems.

9.1 APPLICATION OF MATHEMATICAL MODELS

One of the major benefits associated with the use of mathematical models in corrective measure evaluations relates to the fact that environmental concentrations useful for exposure assessment and risk characterization can be estimated for several locations and time-periods of interest. Since field data frequently are limited and insufficient to accurately and completely characterize a contaminated site and nearby conditions, models can be particularly useful for studying spatial and temporal variabilities, together with potential uncertainties. In addition, sensitivity analyses can be performed by varying specific parameters and then using models to explore the ramifications (as reflected by changes in the model outputs). Models can indeed be used for several purposes in the study of contaminated site problems. In a corrective measure evaluation program, models may be used to determine the general technical feasibility as well as any potential environmental impacts arising from implementation of different remedial actions.

The most common applications of mathematical models in corrective action assessments required for site restoration decisions relate to the following particularly important issues:

- *To understand and predict contaminant fate and transport.* Models can be used to predict contaminant concentrations at receptor point locations or compliance boundaries (e.g. in the prediction of contaminant migration in various environmental compartments, or in the prediction of future concentrations of contaminants at a water supply or compliance boundary well). Models can also be used to predict flow paths and times of travel for contaminants, and therefore to delineate wellhead protection areas.

- *To determine contaminant sources using back-tracking procedures.* Models can be used to help identify contaminant sources by using back-tracking techniques, and working from a given project location. This allows for the equitable allocation of cleanup costs among potentially responsible parties, when several facilities collectively have contributed to contamination at a given project site (such as at well locations beyond a compliance boundary).
- *To screen remedial alternatives.* Models can be used as a screening tool for ranking cleanup alternatives in feasibility studies. For example, screening of alternatives is undertaken to eliminate those remedial actions deemed infeasible due to technical, public health, institutional, and/or cost reasons.
- *Conceptual design of optimal remediation systems.* Models can be used to refine and, in some cases, optimize conceptual designs prior to their implementation (e.g. the design of recharge well fields or basins for efficient injection/infiltration of treated wastewater for irrigation use, and the design and evaluation of remedial schemes which use drains, barrier walls, and caps).
- *Analysis of remedial action alternatives.* Models can be used for the design of monitoring and corrective action plans, or in the simulation of several scenarios during the design of groundwater extraction–injection well networks (e.g. the study of the interaction of surface water bodies, such as small streams, rivers, lakes, with groundwater when the aquifer system is stressed by extraction wells).
- *Performance evaluation of remedial alternatives.* Models can be used to evaluate the expected remedy performance during the feasibility study, so that the anticipated effects of restoration or corrective action can be predicted. For example, models are used to estimate the effects of source-control actions on remediation.

Ultimately, the effective use of models in corrective action assessment and site restoration programs depends greatly on the selection of models most suitable for this purpose. Model selection is dependent on the overall goal of the study, the complexity of the site, and the type of corrective actions being considered. Guidance for effective selection of models in corrective action assessments and site restoration decisions is provided in the literature elsewhere (e.g. CCME, 1994; CHDS, 1990; DOE, 1987; USEPA, 1985, 1987, 1988a, 1988b; Walton, 1984; Zirschy and Harris, 1986).

In general, models usually simulate the response of a simplified version of a complex system. As such, their results are imperfect. Nonetheless, when used in a technically responsible manner, they can provide a very useful basis for making technically sound decisions about a contaminated site problem. They are particularly useful where several alternative scenarios are to be compared. In such cases, all the alternatives are compared on a similar basis; thus, whereas the numerical results of any single alternative may not be exact, the comparative results of showing that one alternative is superior to others will usually be valid.

9.2 SELECTED APPLICATION TOOLS AND LOGISTICS

Corrective measure studies are generally designed to identify and evaluate remedial alternatives that are potentially suitable for addressing well-defined contaminated site problems. Oftentimes, a variety of scientific and analytical tools are employed to assist the decision-maker with the choice of an optimum corrective measure, and indeed in the overall management of the contaminated site problem. A select number of application tools (consisting of scientific models and databases) appropriate for such purposes are enumerated below; a primary communication system to consider in order to obtain further information on the listed softwares would be the on-line service of the Internet, the most widely used international network communication service, otherwise libraries and telephone directories may provide the necessary current contacts. This listing is by no means complete and exhaustive; several other similar logistical tools can indeed be used to support corrective action assessment and corrective measure evaluation programs, in order to arrive at informed decisions on contaminated site problems. In fact, recent years have seen a proliferation of software systems for a variety of environmental management programs. Care must therefore be exercised in the choice of an appropriate tool for specific problems.

9.2.1 AERIS (Aid for Evaluating the Redevelopment of Industrial Sites)

AERIS is an expert system consisting of a multimedia risk assessment model used to generate site-specific cleanup guidelines. It consists of a computer program capable of deriving cleanup guidelines for industrial sites where re-development is being considered.

AERIS serves as a useful remediation model for identifying cleanup objectives. It can be used to identify the factors that are likely to be major contributors to potential exposures and concerns at sites, and those aspects of a re-development scenario with the greatest need for better site-specific information.

AERIS is designed to evaluate situations where the soil had been contaminated sufficiently long enough to establish equilibrium or near-equilibrium conditions. Thus, it is not suitable for evaluating recent spill sites or locations.

Sources of information on AERIS Further information on AERIS may be obtained from the following sources:

● Decommissioning Steering Committee, Canadian Council of Resource and Environment Ministers (CCREM), Canada.
● SENES Consultants Ltd, Richmond Hill, Ontario, Canada.

9.2.2 AIR3D

The American Petroleum Institute (API)'s AIR3D model is a software designed to simulate air flow in the vadose zone during the use of soil vapor extraction (SVE) or soil venting as a site restoration measure. It is a powerful tool used to assist site engineers in the efficient design of vapor extraction systems.

AIR3D is both a deterministic model and an optimization model. Linear programming techniques are incorporated to determine the optimum number and location of venting wells. Optimization can be performed based upon minimizing the number of wells, or on minimizing the cost of installing the system.

Sources of information on AIR3D Further information on AIR3D may be obtained from the following:

- American Petroleum Institute (API), Washington, DC, USA.
- Geraghty & Miller, Inc., Millersville, Maryland, USA.

9.2.3 AIRFLOW/SVE

AIRFLOW/SVE is a comprehensive soil vapor extraction model capable of simulating soil vapor pressure distributions, vapor flow velocities, and multi-component soil vapor concentrations. It can be used to generate graphical displays of contaminant concentration vs. time, and also contaminant mass remaining vs. time for multi-component soil vapors. In addition, the program also contains an on-line database of chemical properties for more than 60 commonly encountered organic contaminants.

AIRFLOW/SVE consists of a combined finite element–finite difference-particle tracking model, developed as a practical tool for the design of SVE systems for site restoration programs. The finite difference formulation of the numerical model allows the flexibility to obtain a reasonably accurate representation of heterogeneous sites, complex boundary conditions, well dimensions, and to some extent residual NAPL contaminant sources. Typical model applications include the simulation of radial-symmetric SVE systems; calculation of effective radius of a SVE system; design of optimal screen position of extraction well; calculation of vapor flow rate for a given vacuum that is created; estimation of cleanup times; and determination of vapor concentrations in vacuum well airstreams.

Source of information on AIRFLOW/SVE Further information on AIRFLOW/SVE may be obtained from the following:

- Waterloo Hydrogeologic, Inc., Waterloo, Ontario, Canada.

9.2.4 AIRTOX (Air Toxics Risk Management Framework)

AIRTOX is a decision analysis model for air toxics risk management. The framework consists of a structural model that relates emissions of air toxics to potential health effects, and a decision tree model that organizes scenarios evaluated by the structural model.

AIRTOX can be used to evaluate the magnitude of health risks to a population, a specific source's contribution to the total health risk, and the cost-effectiveness of current and future emission control measures.

Sources of information on AIRTOX Further information on AIRTOX may be obtained from the following:

- EPRI (Electric Power Research Institute), Palo Alto, California, USA.

9.2.5 API DSS (Exposure and Risk Assessment Decision Support System)

The American Petroleum Institute (API)'s exposure and risk assessment Decision Support System (DSS) is a software system designed to assist environmental professionals in the estimation of human exposures and risk from sites contaminated with petroleum products. It estimates receptor point concentrations by executing fully incorporated unsaturated zone, saturated zone, air emission, air dispersion, and particulate emission models.

The computational modules of the DSS can be implemented in either a deterministic or Monte Carlo mode; the latter is used to quantify the uncertainty in the exposure and risk values that could result from uncertainties in the input parameters.

From physical, chemical, and toxicological property data provided in the DSS databases, risk assessments can be conducted for 16 hydrocarbons, 6 petroleum additives, and 3 metals. The databases can also be expanded to include up to 100 other constituents.

Overall, the DSS is a user-friendly tool that can be used to estimate site-specific exposures and risks; to identify the need for site remediation; to develop and negotiate site-specific cleanup levels with regulatory agencies; and to efficiently and effectively evaluate the effects of uncertainty in the input parameters on estimated risks using Monte Carlo techniques.

Sources of information on API DSS Further information on API DSS may be obtained from the following:

- American Petroleum Institute (API), Washington, DC, USA.
- Geraghty & Miller, Inc., Millersville, Maryland, USA.

9.2.6   GEMS (Graphical Exposure Modeling System)/PCGEMS (Personal Computer Version of the Graphical Exposure Modeling System)

(PC)GEMS is an interactive management tool that allows quick and meaningful analysis of environmental problems. It consists of an interactive computer system for environmental modeling, physico-chemical property estimation, and statistical analysis. The environmental modeling program allows for the simulation of the migration and transformation of chemicals through the air, surface water, soil, and groundwater subsystems.

(PC)GEMS is indeed a complete information management tool designed to help exposure assessment studies. It allows users to estimate chemical properties, assess fate of chemicals in receiving environments, model resulting chemical concentrations, and estimate the resultant human exposure and risk.

Sources of information on (PC)GEMS Further information on (PC)GEMS may be obtained from the following:

- US Environmental Protection Agency (EPA), Research Triangle Park, North Carolina, USA.
- Office of Pesticides and Toxic Substances, Exposure Evaluation Division, US EPA, Washington, DC, USA.
- General Sciences Corporation (GSC), Laurel, Maryland, USA.

9.2.7   HELP (Hydrologic Evaluation of Landfill Performance Model)

HELP is a quasi-two-dimensional deterministic numerical, finite-difference model that computes a daily water budget for a landfill represented as a series of horizontal layers. It models leaching from landfills into the unsaturated soils beneath.

HELP is used for water balance computation, and for the estimation of chemical emissions, and also for leachate quality assessment. It models both organic and inorganic compounds, using rainfall and waste solubility to model the leachate concentrations leaving the landfill.

Sources of information on HELP   Further information on HELP may be obtained from the following:

- US EPA's National Computer Center, Research Triangle Park, North Carolina, USA.
- US Corps of Engineers, Sacramento, California, USA.

9.2.8   HSSM (The Hydrocarbon Spill Screening Model)

HSSM is a screening tool for light nonaqueous phase liquid (LNAPL) impacts to the groundwater table. The model consists of separate modules for addressing LNAPL flow through the vadose zone, LNAPL spreading in the capillary fringe – at the water table – and dissolved LNAPL groundwater transport to potential receptor exposure locations. These modules are based on simplified conceputalizations of the flow and transport phenomena which were used so that the resulting model would be a practical, even if approximate, tool.

The HSSM model is intended for the simulation of subsurface releases of LNAPLs, and is used to estimate the impacts of this type of pollutant on water table aquifers. It offers a simplified approximate analysis for emergency response, for initial phases of site investigations, for facilities siting and permitting, and for underground storage tank programs.

Sources of information on HSSM   Further information on HSSM may be obtained from the following:

- US EPA's Robert S. Kerr Environmental Research Laboratory, Ada, Oklahoma, USA.

9.2.9   IRIS (Integrated Risk Information System) Database

The Integrated Risk Information System (IRIS), prepared and maintained by the Office of Health and Environmental Assessment of the US Environmental Protection Agency (US EPA), is an electronic database containing health risk and regulatory information on several specific chemicals. It is an on-line database of chemical-specific risk information; it is also a primary source of EPA health hazard assessment and related information on several chemicals of environmental concern.

IRIS was originally developed for EPA staff in response to a growing demand for consistent risk information on chemical substances for use in decision-making and regulatory activities. The information in IRIS is accessible to those without extensive training in toxicology, but with some rudimentary knowledge of health and related sciences.

The IRIS database consists of a collection of computer files covering several individual chemicals. These chemical files contain descriptive and numerical information on several subjects, including oral and inhalation reference doses (RfDs) for chronic noncarcinogenic health effects, and oral and inhalation cancer slope factors (SFs) and unit cancer risks (UCRs) for chronic exposures to carcinogens.

IRIS is a tool which provides hazard identification and dose–response assessment information, but does not provide problem-specific information on individual instances of exposure. It is a computerized library of current information that is updated periodically. Combined with specific exposure information, the data in IRIS can be used to characterize the public health risks of a chemical of potential concern under specific scenarios, which can then facilitate the development of effectual corrective action decisions designed to protect public health. The information in IRIS can indeed be used to develop corrective action decision for potentially contaminated sites, such as via the application of risk assessment and risk management procedures.

Sources of information on IRIS Further information on, and access to, IRIS may be obtained from the following:

- IRIS User Support, US EPA, Environmental Criteria and Assessment Office, Cincinnati, Ohio, USA.
- Chemical Information Systems [CIS] (Commercial vendor), Baltimore, Maryland, USA.
- Dialog Information Services, Inc. [DIALOG] (Commercial vendor), Palo Alto, California, USA.
- National Library of Medicine [NLM], Bethesda, Maryland, USA.

9.2.10 IRPTC (International Register of Potentially Toxic Chemicals) Database

In 1972, the United Nations Conference on the Human Environment, held in Stockholm, recommended the setting up of an international registry of data on chemicals likely to enter and damage the environment. Subsequently, in 1974, the Governing Council of the United Nations Environment Programme (UNEP) decided to establish both a chemicals register and a global network for the exchange of information that the register would contain.

In 1976, a central unit for the register, named the International Register of Potentially Toxic Chemicals (IRPTC), was created in Geneva, Switzerland, with the main function of collecting, storing and disseminating data on chemicals, and also to operate a global network for information exchange. IRPTC network partners, the designation assigned to participants outside the central unit, consist of National Correspondents appointed by governments, national and international institutions, national academies of science, industrial research centers and specialized research institutions. Chemicals examined by the IRPTC have been chosen from national and international priority lists. The selection criteria used include the quantity of production and use, the toxicity to humans and ecosystems, persistence in the environment, and the rate of accumulation in living organisms.

IRPTC stores information that would aid in the assessment of the risks and hazards posed by a chemical substance to human health and environment. The major types of information collected include that relating to the behavior of chemicals and information on chemical regulation. Information on the behavior of chemicals is obtained from various sources such as national and international institutions, industries, universities, private databanks, libraries, academic institutions, scientific journals and United Nations bodies such as the International Programme on Chemical Safety (IPCS). Regulatory information on chemicals is largely contributed by IRPTC National Correspondents. Specific criteria are used in the selection of information for entry into the databases. Whenever possible, IRPTC uses data sources cited in the secondary literature produced by national and international panels of experts to maximize reliability and quality. The data are then extracted from the primary literature. Validation is performed prior to data entry and storage on a computer at the United Nations International Computing Center (ICC).

The IPRTC, with its carefully designed database structure, provides a sound model for national and regional data systems. More importantly, it brings consistency to information exchange procedures within the international community. The IPRTC is serving as an essential international tool for chemical hazards assessment, as well as a mechanism for information exchange on several chemicals. The wealth of scientific information contained in the IRPTC can serve as an invaluable database for contaminated site management programs.

Sources of information on IRPTC   Further information on, and access to, IRPTC may be obtained from the following sources:

- National Correspondent to the IRPTC. (Also, following the successful implementation of the IRPTC databases, a number of countries created National Registers of Potentially Toxic Chemicals (NRPTCs), which are completely compatible with the IRPTC system.)

9.2.11   LEADSPREAD

LEADSPREAD provides a methodology for evaluating exposure and the potential for adverse health effects resulting from multi-pathway exposure to inorganic lead in the environment. The method is adapted to a computer spreadsheet. It can be used to determine blood levels associated with multiple pathway exposures to lead at potentially contaminated sites.

LEADSPREAD basically consists of a mathematical model for estimating blood lead concentrations as a result of contacts with lead-contaminated environmental media. A distributional approach is used, allowing estimation of various percentiles of blood lead concentration associated with a given set of inputs.

Sources of information on LEADSPREAD   Further information on LEADSPREAD may be obtained from the following:

- Office of Scientific Affairs, Department of Toxic Substances Control (DTSC), California EPA, Sacramento, California, USA.

9.2.12 MULTIMED (Multimedia Exposure Assessment Model)

MULTIMED is a computer model for simulating the transport and transformation of contaminants released from a hazardous waste disposal facility into the multimedia environment. The MULTIMED model simulates releases into air and soil – including the unsaturated (vadose) and saturated zones, and possible interception of the subsurface contaminant plume by a surface stream. It further simulates movement through the air, soil, groundwater and surface water media to contact humans and other potentially affected receptors. Uncertainties in parameter values used in the model are quantified using Monte Carlo simulation techniques.

MULTIMED is typically used to simulate the movement of contaminants leaching from a waste disposal facility. It is intended for general exposure and risk assessments of waste facilities, and for the analyses of the impacts of engineering and management controls.

Sources of information on MULTIMED   Further information on MULTIMED may be obtained from the following:

- Environmental Research Laboratory, Office of Research & Development, US EPA, Athens, Georgia, USA.

9.2.13 RAPS (Remedial Action Priority System)

RAPS was developed for use by the US Department of Energy (DOE), to aid in the setting of priorities for the investigation and possible cleanup of chemical and radioactive waste disposal sites. It is intended to be used in a comparative rather than predictive mode.

The RAPS methodology considers four major pathways of contaminant migration: groundwater, surface water, overland flow, and atmospheric. Estimated concentrations in the air, soil, sediments, and water media are used to assess exposure to neighboring populations. The estimated environmental concentrations form the basis of subsequent human exposure calculation, and the determination of a 'Hazard Potential Index' (HPI). The RAPS methodology is not truly multimedia since it is based on use of independent modules which do not interact spatially or temporally; that is, transfer of pollutant is in one direction only.

Sources of information on RAPS   Further information on RAPS may be obtained from the following:

- Battelle Pacific Northwest Laboratory, Richland, Washington, USA.

9.2.14 RBCA (Risk-Based Corrective Action) Spreadsheet System

The RBCA (risk-based corrective action) spreadsheet system/tool kit is a complete step-by-step package for the calculation of site-specific risk-based soil and groundwater cleanup goals, which will then facilitate the development of site remediation plans. The system includes fate and transport models for major and significant exposure pathways (i.e. air, groundwater, and soil), together with an integrated chemical/toxicological library of several chemical compounds (i.e. over 80, and also expandable by the user).

The RBCA process allows for the calculation of baseline risks and cleanup standards, as well as for remedy selection and compliance monitoring at petroleum release sites. The user simply provides site-specific data to determine exposure concentrations, average daily intakes, baseline risk levels, and risk-based cleanup levels.

Sources of information on RBCA Further information on the RBCA tool kit/ spreadsheet system may be obtained from the following:

- Groundwater Services, Inc., Houston, Texas, USA.
- Environmental Systems & Technologies, Inc., Blacksburg, Virginia, USA.

9.2.15 ReOpt

ReOpt is a remediation software that can be used in the selection of suitable technologies for the cleanup of contaminated sites. It speeds up site cleanup decisions, because the ReOpt software enables a user to, quickly and easily, review a variety of remediation options and determine their effectiveness for the particular site under investigation.

ReOpt contains information about technologies that might potentially be used for cleanup at contaminated sites; auxiliary information about possible hazardous or radioactive contaminants at such sites; and selected pertinent regulations that govern disposal of wastes containing these contaminants. The user specifies a series of conditions, and ReOpt provides a short list of cleanup technology choices specific to the particular situation. The technology selection is based on site characteristics and cleanup strategy.

The ReOpt software enables engineers and planners involved in environmental restoration efforts to quickly identify potentially applicable environmental restoration technologies and to access corresponding information required to select cleanup activities for contaminated sites. The analyst can automatically select potentially appropriate technologies by simply specifying the contaminants or contaminated medium of interest. The analyst can also select any technology and then review the technical description of the relevant process with accompanying schematic diagrams, as well as examine the technical and regulatory constraints which govern the technology.

Sources of information on ReOpt Further information on ReOpt may be obtained from the following:

- Sierra Geophysics, Inc., Seattle, Washington, USA.
- Battelle Pacific Northwest Laboratory, Richland, Washington, USA.

9.2.16 RISC (Risk Identification of Soil Contamination)

RISC (risk identification of soil contamination) is a knowledge-based framework for risk identification and evaluation of sites with contaminated soils. It consists of computer modules that facilitate site investigations, risk analyses, and priority-ranking for former industrial facilities.

The RISC framework uses expert information on the fate and behavior of contaminants in soil systems to predict potential risks to human health and the environment, that could result from contaminated site problems. Dutch, English, and German versions of this computer model system are available.

Sources of information on RISC Further information on the RISC computer model system may be obtained from the following:

• Van Hall Institute, Groningen, The Netherlands.

9.2.17 RISK*ASSISTANT

RISK*ASSISTANT provides an array of analytical tools, databases, and information-handling capabilities for risk assessment. It has the ability to tailor exposure and risk assessments to local conditions.

The RISK*ASSISTANT software is designed to assist the user in rapidly evaluating exposures and human health risks from chemicals in the environment at a particular site. The user need only provide measurements or estimates of the concentrations of chemicals in the air, surface water, groundwater, soil, sediment, and/or biota.

*Sources of information on RISK*ASSISTANT* Further information on RISK* ASSISTANT may be obtained from the following:

• Hampshire Research Institute, Alexandria, Virginia, USA.
• US EPA, Research Triangle Park, North Carolina, USA.
• California EPA, Sacramento, California, USA.
• New Jersey Department of Environmental Protection, Trenton, New Jersey, USA.

9.2.18 RISKPRO

RISKPRO is a complete software system designed to predict the environmental risks and effects of a wide range of human health-threatening situations. It consists of a multimedia/multipathway environmental pollution modeling system, providing for modeling tools to predict exposure from pollutants in the air, soil and water.

RISKPRO is used to evaluate receptor exposures and risks from environmental contaminants. It graphically represents its results through maps, bar charts, wind-rose diagrams, isopleth diagrams, pie charts, and distributional charts. Its mapping capabilities can also allow the user to create custom maps showing data and locations of environmental contaminant plumes.

Sources of information on RISKPRO Further information on RISKPRO may be obtained from the following:

• General Sciences Corporation (GSC), Laurel, Maryland, USA.

9.2.19 SITES (The Contaminated Sites Risk Management System)

SITES is a flexible interactive PC computerized decision-support tool for organizing relevant information needed to conduct risk management analyses for contaminated sites. It has the dimensionality to model multiple chemicals, pathways, population groups, health effects, and remedial actions. The model uses information from diverse sources, such as site investigations, transport and fate modeling, behavioral and exposure estimates, and toxicology.

SITES is indeed a computer-based integrating framework used to help evaluate and compare site investigation and remedial action alternatives in terms of health and environmental effects and total economic costs/impacts. The user completely defines the scope of the analyses. Both deterministic and probabilistic analyses are possible. The decision-tree structure in SITES allows for explicit examination of key uncertainties and the efficient evaluation of numerous scenarios. The model's design and computer implementation facilitates quick and extensive sensitivity analyses.

Sources of information on SITES   Further information on SITES may be obtained from the following:

• EPRI (Electric Power Research Institute), Palo Alto, California, USA.

9.2.20   SUTRA (Saturated–Unsaturated Transport Model)

SUTRA is a two-dimensional numerical, finite-element, and integrated finite-element solution technique. It is a solute transport simulation model, which may be used to model natural or human-induced chemical species transport, including processes of solute sorption, production, and decay.

SUTRA may be applied to the analysis of groundwater contaminant transport and aquifer restoration designs. It predicts fluid movement and the transport of either energy or dissolved substances in a subsurface environment.

Sources of information on SUTRA   Further information on SUTRA may be obtained from the following:

• USGS (US Geological Survey), Water Resources Department, Reston, Virginia, USA.

9.2.21   WET (Wastes–Environments–Technologies Model)

WET is a risk/cost policy model that establishes a system to allow users to investigate how tradeoffs of costs and risks can be made among wastes, environments, and technologies in order to arrive at feasible regulatory options. The system assesses waste streams in terms of likelihood and severity of human exposure to their hazardous constituents, and then models their behavior in three media (viz.: air, surface water and groundwater).

WET is typically used to assist policymakers in identifying cost-effective options that minimize risks to health and the environment.

Sources of information on WET   Further information on WET may be obtained from the following:

• Office of Health and Environmental Assessment, US EPA, Washington, DC, USA.

9.3 REFERENCES

CCME (Canadian Council of Ministers of the Environment) (1994). *Subsurface Assessment Handbook for Contaminated Sites*. Canadian Council of Ministers of the Environment (CCME), The National Contaminated Sites Remediation Program (NCSRP), Report No. CCME-EPC-NCSRP-48E (March 1994), Ottawa, Ontario, Canada.

CDHS (California Department of Health Services) (1990). *Scientific and Technical Standards for Hazardous Waste Sites*. Prepared by the California Department of Health Services, Toxic Substances Control Program, Technical Services Branch, Sacramento, California.

DOE (US Department of Energy) (1987). *The Remedial Action Priority System (RAPS): Mathematical Formulations*. US Department of Energy, Office of Environment, Safety & Health, Washington, DC.

USEPA (US Environmental Protection Agency) (1985). *Modeling Remedial Actions at Uncontrolled Hazardous Waste Sites*. EPA/540/2-85/001 (April 1985). Office of Emergency and Remedial Response, Washington, DC.

USEPA (US Environmental Protection Agency) (1987). *Selection Criterion for Mathematical Models Used in Exposure Assessments: Surface Water Models*. EPA-600/8-87/042. Office of Health and Environmental Assessment, Washington, DC.

USEPA (US Environmental Protection Agency). (1988a). *Selection Criteria for Mathematical Models Used in Exposure Assessments: Ground-Water Models*. EPA-600/8-88/075. Office of Health and Environmental Assessment, Washington, DC.

USEPA (US Environmental Protection Agency) (1988b). *Superfund Exposure Assessment Manual*. Report No. EPA/540/1-88/001, OSWER Directive 9285.5-1, USEPA, Office of Remedial Response, Washington, DC.

Walton, W.C. (1984). *Practical Aspects of Ground Water Modeling*. National Water Well Association.

Zirschy, J.H. and D.J. Harris (1986). Geostatistical analysis of hazardous waste site data. *ASCE J. Environ. Engnr.* **112**(4).

Chapter Ten

Evaluation of site restoration options

Contaminated site restoration programs generally involve the use of containment or cleanup strategies. Containment strategies have the goal of preventing further migration of mobile contaminants, by controlling contaminant plume movements within a specified area and time-frame; cleanup strategies have the goal of removing contaminants or contaminated media in a specified area until acceptable concentration levels are attained. The development, screening, and selection of the preferred remedial action alternatives involves identifying a range of remediation options that will ensure adequate protection of public health and the environment. Depending on the site-specific circumstances, the selected remedial option may result in the complete elimination or destruction of the contaminants of concern that are present at a project site, the reduction of contaminant concentrations to 'acceptable' risk-based levels, and/or the prevention of exposure to the contaminants of concern via engineering or institutional controls.

Typically, once a list of remedial alternatives is developed, these alternatives are analyzed in detail so that the most appropriate option for the site-specific problem can be selected. The analyses usually involve an initial screening, followed by a detailed evaluation. The initial screening of alternatives is designed to eliminate alternatives which are clearly inappropriate to the given situation or are clearly inferior to other alternatives; the alternatives which remain after the initial screening are subjected to more detailed evaluation. Based on the results of the detailed analysis, the appropriate remedial alternative(s) can then be selected. The selection of a particular type of remedial option depends on the type of contaminants involved, cleanup requirements, cost-effectiveness, practicability, general site conditions and accessibility, and applicable local regulations that must be met in the site restoration process.

10.1 DEVELOPMENT AND SCREENING OF ALTERNATIVE SITE RESTORATION OPTIONS

In the process of developing contaminated site remediation alternatives, information on the nature and extent of contamination, applicable local environmental regulations, contaminant fate and transport properties, and the toxicity of contaminants are used to guide decisions made about the potentially feasible and appropriate remedial options. Subsequently, the remedial action alternatives and associated technologies are screened to identify those that will likely be effective for the contaminants and media of interest at the specific project site. The broad groups of remedial alternatives generally screened during corrective measure assessments for contaminated site problems include: containment, removal, and treatment of contaminated materials. Site characterization data are used to identify the general approach, or combination of approaches, which is likely to be most effective in addressing each impacted environmental matrix at a contaminated site.

The development of remedial alternatives   The development of remedial alternatives for contaminated site problems involves compiling a limited number of site restoration options for source control and/or remedial action. Typically, the development of a remedial action alternative will consist of several activities, including the following (Pratt, 1993; USEPA, 1988a, 1988b):

- Establish site restoration objectives and remediation goals.
- Identify potential treatment technologies and containment or disposal requirements for the contaminants of concern (which will satisfy the site restoration objectives and remediation goals).
- Determine process options and general response actions (which will satisfy the site restoration objectives and remediation goals).
- Identify areas of impacted media (to which general response actions might be required).
- Pre-screen remedial technologies and process options based on their effectiveness, implementability, and cost.
- Assemble technologies and their associated containment or disposal requirements into alternatives for the contaminated media.
- Identify regulatory limits or cleanup criteria, and compare with removal efficiencies of remedial techniques.
- Determine the area over which cleanup levels will be achieved for the contaminated site, encompassing the area outside the site boundary and up to the boundary of contaminant plume.
- Estimate the restoration time-frame, comprising of the period of time required to achieve selected cleanup levels at all locations within the area designated for site restoration.
- Compile and group remediation technologies and treatment processes into appropriate remedial alternatives.

The remedial alternatives are usually developed after a site characterization program in order to specify the area designated for site restoration, the restoration time-frame, the cleanup levels, and the feasible remediation techniques.

The screening of remedial alternatives   In most situations, several potentially feasible remedial options are developed early on in the site restoration evaluation process. Consequently, it becomes necessary to screen out some of the available options, in order to reduce the number of alternatives that will be analyzed in detail. The screening process, usually done on a general basis and with limited effort, involves evaluating alternatives with respect to their effectiveness, implementability, and cost. In fact, because the screening process addresses approaches to site restoration rather than specific remedial technologies, the evaluation is more qualitative (rather than being quantitative). However, the screening analysis uses the quantitative site characterization data to recommend an approach to the site restoration program.

In general, it is important to consider as many alternatives as possible during the screening of remedial action measures. This will ensure that the most cost-effective technique is not excluded from consideration. On the other hand, it is impractical or uneconomical to conduct extensive and detailed evaluation of every remedial alternative during the planning and preliminary design stages. Thus, the first step is to determine the potentially feasible alternatives that can be evaluated further, based on technical and economic factors.

10.2   THE DETAILED ANALYSIS OF ALTERNATIVE SITE RESTORATION OPTIONS

The detailed analysis of remedial alternatives is conducted with the principal objective of providing decision-makers with sufficient information to compare alternatives in a technically justifiable and socio-economically acceptable manner. It follows the development and screening of feasible alternatives and precedes the actual selection of a remedy. Following the screening of remedial alternatives, a detailed analysis is conducted to identify the remedial technology most likely to be successful, from among the remedial approaches previously compiled during the screening analysis.

The detailed evaluation of the applicable alternative technologies being considered for a contaminated site problem will typically incorporate information on the successful application of the technology under similar site conditions, total project cost, attainment of acceptable risk reduction, project duration, and the manageability of project data requirements. In general, the detailed evaluation of site restoration options is guided by the application of the following specific set of evaluation criteria: protection of human health and the environment; short-term effectiveness; long-term effectiveness; compliance with regulatory standards; reduction of toxicity, mobility, or volume; technical and administrative implementability; benefit–cost ratios; and regulatory and community acceptance (USEPA, 1988a, 1989). The process helps determine the respective strengths and weaknesses of alternative remedial measures, and to identify the key tradeoffs that must be balanced for a contaminated site problem. The results of the detailed evaluations will comprise of a recommended technology or combination of technologies to restore each impacted medium posing 'unacceptable' risks.

The risk evaluation of remedial alternatives   The evaluation of both short-term and long-term risks is an important part of the detailed analyses of site restoration options. In fact, risk assessment plays a very important role in the development of remedial action objectives for contaminated sites; in the identification of feasible remedies that meet the remediation objectives; and in the selection of an optimum remedial alternative. The processes involved in the risk evaluation of remedial alternatives consist of the same general steps as a baseline risk assessment (discussed in Chapter 6), except that the baseline risk assessment typically is more refined than the risk comparison of remedial alternatives. The difference between the site risks in the absence of remedial action (i.e. the baseline risk) and the risks associated with a remedial alternative will generally help define the net benefits associated with a given remedial option. Overall, it is crucial to ensure that the projected risks posed by a remedial option do not offset any benefits associated with reducing site contamination in order to achieve an established site restoration or risk reduction goal.

Risk assessment techniques can indeed be used to quantify the human health risks and environmental hazards created by implementing specific remedial options at contaminated sites. These procedures can help determine whether a particular remedial alternative will pose unacceptable risks following implementation, and to determine the specific remedial alternatives that will create the least risk with respect to the cleanup goals or remedial action objectives for the site. Consequently, risk assessment tools can be used to aid in the process of selecting among remedial options for contaminated sites.

10.3   THE SELECTION OF A SITE RESTORATION OPTION

The principal remediation options that seem to have found widespread applications in the management of contaminated site problems have usually involved the removal of contaminated material from the case site for disposal elsewhere; retention and isolation of contaminated material on-site using an appropriate form of cover, barrier, or encapsulation system; physical, chemical or biological treatment to eliminate or immobilize the contaminants; and/or lowering of the contaminant concentrations by diluting the contaminated medium with clean material (Cairney, 1993). Ultimately, the selection of an appropriate site restoration strategy for a contaminated site problem depends on a careful assessment of both short- and long-term risks posed by the site, which ensures that the selected remedy will satisfy the following pertinent conditions:

- Protectiveness of human health and the environment.
- Attainment of media cleanup standards and/or site restoration goals.
- Control of the source(s) of release in order to reduce or eliminate, to the extent practicable, further releases that may pose a threat to human health or the environment.

In general, there is no single remedial technique that is best for all types of contamination and for all site conditions. In fact, successful remediation efforts may have to rely on proper marriages between remediation technologies. Also, depending on the circumstances, alternative remedial strategies may be developed for a contaminated site as a whole, for individual areas of the site (e.g. lagoons, waste stockpiles, etc.) or for a specific impacted medium (e.g. groundwater, soils, etc.).

The general criteria for the selection and implementation of site restoration options for contaminated site problems are discussed below; further details can be found elsewhere in the literature (e.g. Cairney, 1993; Calabrese and Kostecki, 1991; Nyer, 1993; OBG, 1988; Pratt, 1993; Sims et al., 1986; USEPA, 1984, 1985, 1988b). Overall, the selection of one remedial approach over another will depend on such factors as the ease of implementation with respect to technical feasibility and regulatory compliance; brevity of project duration; effectiveness of reducing contamination and risk to 'acceptable' levels; and attainment of reasonable risk reduction within a justifiable cost–benefit framework.

10.3.1   Remedy Selection Criteria

Site restoration options considered for contaminated sites will generally be evaluated based on several criteria. The particularly important remedy selection decision factors that should be considered in the evaluation of site restoration options include the following (Charbeneau et al., 1992; USEPA, 1988a, 1988b):

- Overall protection of human health and the environment
- Compliance with applicable laws and regulations
- Short- and long-term effectiveness, and permanence of corrective actions
- Potential short- and long-term liability issues
- Type and amount of contaminated media
- Ease of implementation
- Cost-effectiveness and cost-efficiency of plans
- Reduction of toxicity, mobility or volume of contaminated materials
- Regulatory and community acceptance of the program
- Future land-uses.

Ultimately, the selected remedy will be the alternative found to provide the best balance of tradeoffs among alternatives in terms of these evaluation criteria. This will generally satisfy several important requirements, such as providing the lowest cost alternative that is technologically feasible and reliable, and which effectively mitigates and minimizes environmental damage, as well as provides adequate protection of public health, welfare, or the environment. Any remedies not meeting these criteria are eliminated from further consideration as a preferred alternative.

In general, the remedial alternative selected following the detailed evaluation should attain or exceed pertinent regulatory standards that apply to the site, and should also realize sustained effectiveness. Indeed, a number of other extraneous but important site-specific features may also affect the selection of the ultimate corrective measure; these include several site characteristics pertaining to surface features, subsurface conditions, populations potentially at risk, climate, adjacent land-uses, cultural and social situations, and local regulatory climate.

10.3.2   Choosing between Site Restoration Options

The selection of a specific remedial alternative to address a contaminated site problem depends, to a great extent, on the required cleanup criteria and site restoration goals established for the site. Once the cleanup criteria have been determined, a variety of remedial techniques can be evaluated for containing and treating impacted media associated with a contaminated site problem. The remediation approach may comprise of the use of physical containment techniques and/or physical, chemical, and biological treatment processes. In a typical situation, the remedial action will include the use of techniques to contain the contamination plume, and to recover and treat the impacted matrices.

Overall, successful and long-term site restoration requires adequate mitigation strategies to remove contaminant source(s). Source control technologies which involve treatment of contaminated materials, or which otherwise do not rely on containment structures or systems to prevent future releases, should be strongly preferred to those that offer temporary, or less reliable, controls.

10.3.3   Performance Evaluation of Remedial Techniques

The principal objective of a contaminated site restoration program is to protect human health, the environment, and public and private properties in the vicinity of the impacted site. Contaminated site risks are generally eliminated, reduced, or controlled via treatment, engineering measures, or institutional controls. Usually, the amount of reduction in toxicity, mobility, or volume offered by treatment processes gives a measure of the anticipated performance of the remedial technique employed at a project site. This measure of performance becomes a particularly important consideration in the selection and implementation of a specific remedial action as opposed to the alternative(s) that are eliminated.

The performance goal of remedial alternatives may be evaluated based on a cleanup criterion and a time period for the restoration of the contaminated site. The favored alternatives are compared based on the tradeoffs between the time to attain an 'acceptable' cleanup level and the costs associated with the remedial actions. It should be recognized, however, that the complexities in the contaminant fate and transport mechanisms at contaminated sites often make it difficult to predict the performance of site restoration actions with a high enough degree of accuracy.

10.4 REFERENCES

Cairney, T. (ed.) (1993). *Contaminated Land (Problems and Solutions)*. Blackie Academic & Professional, Glasgow/Chapman and Hall, London/Lewis Publishers, Boca Raton, Florida.

Calabrese, E.J. and P.T. Kostecki (eds) (1991). *Hydrocarbon Contaminated Soils*, Vol. 1. Lewis Publishers, Inc., Chelsea, Michigan.

Charbeneau, R.J., P.B. Bedient, and R.C. Loehr (eds) (1992). *Groundwater Remediation. Water Quality Management Library*, Vol. 8. Technomic Publishing Co., Inc., Lancaster, Pennsylvania.

Nyer, E.K. (1993). *Practical Techniques for Groundwater and Soil Remediation*. Lewis Publishers, Boca Raton, Florida.

OBG (O'Brien & Gere Engineers, Inc.) (1988). *Hazardous Waste Site Remediation: The Engineer's Perspective*. Van Nostrand Reinhold, New York.

Pratt, M. (ed.) (1993). *Remedial Processes for Contaminated Land*. Institution of Chemical Engineers, Warwickshire, UK.

Sims, R. et al. (1986). *Contaminated Surface Soils In-Place Treatment Techniques*. Noyes Publications, Park Ridge, New Jersey.

USEPA (US Environmental Protection Agency) (1984). *Review of In-place Treatment Techniques for Contaminated Surface Soils*, Vols 1 and 2. US Environmental Protection Agency, Hazardous Waste Engineering Research Laboratory, Cincinnati, OH. EPA-540/2-84-003a and b.

USEPA (US Environmental Protection Agency) (1985). *Modeling Remedial Actions at Uncontrolled Hazardous Waste Sites*. EPA/540/2-85/001 (April 1985). Office of Emergency and Remedial Response, Washington, DC.

USEPA (US Environmental Protection Agency) (1988a). *Guidance for Conducting Remedial Investigations and Feasibility Studies Under CERCLA*. EPA/540/G-89/004. OSWER Directive 9355.3-01, Office of Emergency and Remedial Response, Washington, DC.

USEPA (US Environmental Protection Agency) (1988b). *Guidance on Remedial Actions for Contaminated Ground Water at Superfund Sites*. Office of Emergency and Remedial Response, Washington, DC. EPA/540/G-88/003.

USEPA (US Environmental Protection Agency) (1989). *Risk Assessment Guidance for Superfund*. Vol. 1 – *Human Health Evaluation Manual* (Part A). EPA/540/1-89/002. Office of Emergency and Remedial Response, Washington, DC.

Chapter Eleven

Development of a site restoration plan for a contaminated site problem: an illustrative example

The purpose of this chapter is to present a procedural illustration of the types of evaluation required for the development of a site restoration program, as part of a decommissioning or closure plan for an abandoned industrial facility. The case site, owned by PLC Limited, is located within an industrial estate in the outskirts of London. This hypothetical facility has been used for a multitude of operations – including machine components cleaning, electroplating, sandblasting, painting, and vehicle maintenance. An environmental site assessment conducted for the PLC facility indicated a high degree of soil and groundwater contamination within the site boundaries. This occurrence is the result of the past site activities. Based on current zoning plans, it is anticipated that this land parcel could be used for a variety of commercial developments in the near future. This illustrative problem consists of the development of a corrective action response plan to address the contamination problem encountered at this site.

11.1 INTRODUCTION AND BACKGROUND

The former industrial facility, located in an industrially zoned area in the outskirts of metropolitan London, operated for over three decades before being permanently closed. Site facilities include a main plant building, office buildings, storage tanks, and post-closure areas (that consist of surface impoundments for wastewater treatment operations and sludge ponds). Past operations at the plant required the storage of raw materials in above-ground tanks, the distribution of raw materials in pipelines, and the storage of chemicals, fuels and waste materials in USTs. Historical uses of the site included machine component cleaning (in which chlorinated hydrocarbon-based solvents were used) and electroplating (for which major associated chemicals included cadmium, nickel, and chromium). Other significant activities included sandblasting of unpainted metal parts, painting, and vehicle maintenance.

Due to the sandblasting activities, incidental spillage during materials handling, and possible leakage of underground storage and distribution systems, soils and ground-water underlying the PLC plant site have been significantly impacted; this is the result of releases of chemical materials that were used in the industrial processes and related activities carried out at this facility. Preliminary remedial activities have already been implemented to remove buried drums and storage tanks, and to remove soil materials from some of the most heavily contaminated areas.

The principal goal of any comprehensive corrective action program for the PLC site would be to prevent contaminant migration from the site to potential receptors, and therefore prevent the endangerment of human health and the environment at and in the vicinity of the site. The overall objective of the corrective measure assessment for the PLC facility is to determine the type of remedial systems necessary to abate potential risks posed by the site.

11.2 KEY ENVIRONMENTAL CONCERNS

It is apparent that releases at the PLC site have caused significant soil and groundwater contamination beneath this industrial facility. In this hypothetical example, the key environmental issues affecting the development of a site restoration strategy relate to the following:

- Identification of the possible site-activity-related contaminants associated with the site.
- Screening for the chemicals of potential concern to human health and the environment.
- Estimation of the chemical concentrations in the impacted media of significant concern.
- Determination of the populations potentially at risk from site contaminants.
- Identification of site-specific exposure scenarios that give an adequate conceptual representation of the site.
- Characterization of the potential risks associated with the site.
- Development of site-specific cleanup criteria for the impacted matrices at the site.

Based on the type of exposure scenarios identified for this environmental setting, a cleanup strategy can be developed to adequately control potential risks to human health and the environment posed by this hypothetical site.

11.3 A DIAGNOSTIC RISK ASSESSMENT

This section consists of a baseline risk assessment for the inactive site that previously housed the PLC facility. Under the current decommissioning program, it is expected that the land parcel at the PLC facility could be zoned for a variety of commercial developments in the near future. The development of a site closure or re-development plan should therefore incorporate a diagnostic risk assessment that addresses potential impacts under all realistically feasible site uses and conditions.

Identification of site contaminants and the screening for chemicals of potential concern Soils and groundwater present at the PLC site appear to have been significantly impacted as a result of releases of chemical materials that were used in the past site activities. A site characterization program has been undertaken to define the nature and extent of the soil and groundwater contamination within the site boundary. Chemicals found in soils and groundwater at the PLC site consist of both organic and inorganic constituents, as summarized in Tables 11.1 and 11.2 (developed from the complete laboratory data package of environmental sampling results by using

Figure 11.1). The summary of analytical results is reported in these tables, together with the naturally-occurring background threshold values, where background concentrations are available. The background levels are used as a screening indicator of possible media contamination that may be the result of past site activities.

A listed site contaminant is considered to be a chemical of potential concern if it may be attributed to the site if it could result in adverse effects to populations potentially at risk when such receptors are exposed to the particular compound. To obtain the chemicals of potential concern for the anticipated future use of the site, Figure 11.1 is again used to screen the site contaminants in order to arrive at the target chemicals listed in Table 11.3.

A conceptualization of the site-specific exposure scenarios for risk characterization Table 11.4 provides a summary of the anticipated migration and exposure pathways at the PLC site. Exposure scenarios, involving the three different population groups, that are selected as being complete and significant are evaluated further using the methods of approach previously discussed in Chapters 6 and 7. The maximum concentrations of the target chemicals in the environmental samples are used as the exposure point concentrations or their equivalents in this evaluation. Also, case-specific exposure parameters obtained from the literature (viz. DTSC, 1994; OSA, 1993; USEPA, 1989a, 1989b, 1991, 1992) are used in the modeling effort. Toxicity values for the risk characterization pertain to those found in recent toxicological databases. Calculation of potential carcinogenic risks and noncarcinogenic hazards under the existing conditions at the PLC site are performed for the three different population groups identified in the conceptual site model (Table 11.4); for the noncarcinogenic effects, it is assumed for the sake of simplicity, that all the target chemicals have the same physiologic endpoint.

An evaluation of the potential impacts of the chemicals of potential concern present at the PLC site, under the various exposure scenarios, follows.

Risk characterization associated with an on-site worker Table 11.5 consists of an evaluation of the potential risks associated with a nearby and/or on-site worker (following the re-development of the site for commercial activities) being exposed to the chemicals of potential concern at the PLC site, assuming the contaminated soils remain in place. It is assumed that potential receptors may be exposed via inhalation of airborne contamination (consisting predominantly of particulate emissions from fugitive dust), through the incidental ingestion of contaminated soils, and by dermal contact with the contaminated soils at the site. Case-specific exposure parameters used in this evaluation conservatively assume that the on-site worker will be exposed at a frequency of 250 days per year over a 25-year period. Additional parameters include using a soil ingestion rate of 50 mg/day and airborne particulate emission rate of 50 μg/m^3. Other default exposure parameters indicated in the literature (viz. DTSC, 1994; OSA, 1993; USEPA, 1989a, 1989b, 1991, 1992) were used for the calculations shown in this spreadsheet. Based on this scenario, it is apparent that potential risks to an on-site worker at the PLC site could exceed an assumed benchmark risk level of 10^{-6}. The noncarcinogenic hazard index, however, is within the reference index of unity. The sole 'risk driver' in this case is arsenic.

Risk characterization associated with a site construction worker Table 11.6 consists of an evaluation of the potential risks associated with a construction worker being

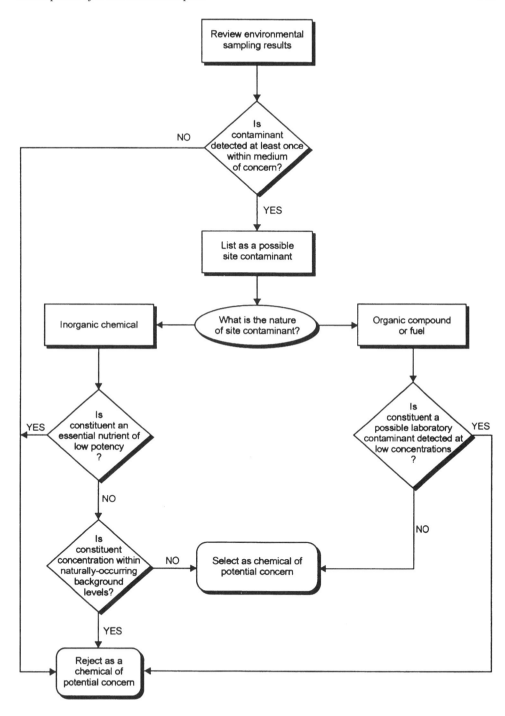

Figure 11.1 Decision process used to screen for the chemicals of potential concern in the impacted environmental matrices

Table 11.1   Preliminary list of possible site contaminants in soils at the PLC site

Possible site contaminant	Important synonyms, trade names, or chemical formula	Naturally-occurring background level (mg/kg)	Maximum soil concentration (mg/kg)
INORGANIC CHEMICALS			
Aluminum	Al	16 400	10 400
Antimony	Sb	0.3	1.1
Arsenic	As	10	12
Barium	Ba	98	81.1
Beryllium	Be	0.7	0.33
Cadmium	Cd	1.0	2.6
Calcium	Ca	3 350	8 350
Chromium	Cr	28	57.3
Cobalt	Co	4.0	64.6
Copper	Cu	46	47.3
Iron	Fe	13 500	22 900
Lead	Pb	22	15
Magnesium	Mg	3 900	2 420
Manganese	Mn	523	170
Mercury	Hg	0.2	1.1
Molybdenum	Mo	3.0	4.8
Nickel	Ni	13	19
Potassium	K	2 610	2 200
Selenium	Se	0.2	6.1
Silver	Ag	0.2	0.55
Sodium	Na	453	588
Thallium	Tl	0.4	0.415
Vanadium	V	58	34.1
Zinc	Zn	107	432
ORGANIC COMPOUNDS			
Trichloroethene	TCE	not available	0.710
cis-1,2-Dichloroethene	*cis*-1,2-DCE	not available	0.007

exposed to the chemicals of potential concern at the PLC site, assuming no personal protection from the contaminated soils during site re-development activities. It is assumed that potential receptors may be exposed via inhalation of airborne contamination (consisting predominantly of particulate emissions from fugitive dust), through the incidental ingestion of contaminated soils, and by dermal contact with the contaminated soils at the site. Case-specific exposure parameters used in this evaluation conservatively assume that the construction worker will be exposed at a frequency of 250 days per year over a one-year period. Additional parameters include using a soil ingestion rate of 480 mg/day and airborne particulate emission rate of 1000 μg/m^3. Other default exposure parameters indicated in the literature (viz. DTSC, 1994; OSA, 1993; USEPA, 1989a, 1989b, 1991, 1992) were used for the calculations shown in this spreadsheet. Based on this scenario, it is apparent that potential risks to a construction

Table 11.2 Preliminary list of possible site contaminants in groundwater at the PLC site

Possible site contaminant	Important synonyms, trade names, or chemical formula	Naturally-occurring background level (μg/L)	Maximum water concentration (μg/L)
INORGANIC CHEMICALS			
Aluminum	Al	1 200	1 000
Antimony	Sb	10	5.9
Arsenic	As	7	5.0
Barium	Ba	276	133
Beryllium	Be	4.0	1.0
Calcium	Ca	197 000	54 300
Chromium	Cr	20	15
Cobalt	Co	13	6.3
Copper	Cu	58	22
Iron	Fe	3 530	3 000
Magnesium	Mg	119 000	37 800
Manganese	Mn	971	1 390
Molybdenum	Mo	6	6.5
Nickel	Ni	490	21
Potassium	K	13 300	6 550
Silver	Ag	12	4.0
Sodium	Na	420 000	124 000
Thallium	Tl	1.0	1.5
Vanadium	V	28	24
Zinc	Zn	80	70
ORGANIC COMPOUNDS			
Trichloroethene	TCE	not available	939
cis-1,2-Dichloroethene	*cis*-1,2-DCE	not available	118

worker at the PLC site could marginally exceed an assumed benchmark risk level of 10^{-6}. The noncarcinogenic hazard index also marginally exceeds the reference index of 1. The major 'risk drivers' are arsenic and cobalt.

Risk characterization for a downgradient residential population exposure to groundwater Table 11.7 consists of an evaluation of the potential risks associated with a hypothetical downgradient population exposure to impacted groundwater that originates from the PLC site, assuming the contaminated water is not treated before going into a public water supply system. It is assumed that potential receptors may be exposed through the inhalation of volatiles during domestic usage of contaminated water, from the ingestion of contaminated water and by dermal contact to contaminated waters. Default exposure parameters indicated in the literature (viz. DTSC, 1994; OSA, 1993; USEPA, 1989a, 1989b, 1991, 1992) were used for the calculations presented in this spreadsheet. Based on this scenario, it is apparent that the generally acceptable benchmark risk level of 10^{-6} and the reference hazard index of 1

Table 11.3 Summary of the chemicals of potential concern at the PLC site

Chemical of potential concern in soils	Important synonyms or trade names, or chemical formula	Maximum soil concentration (mg/kg)	Chemical of potential concern in groundwater	Important synonyms or trade names, or chemical formula	Maximum groundwater concentration (µg/L)
INORGANIC CHEMICALS			INORGANIC CHEMICALS		
Antimony	Sb	1.1	Manganese	Mn	1 390
Arsenic	As	12.0	Molybdenum	Mo	6.5
Cadmium	Cd	2.6	Thallium	Tl	1.5
Chromium	Cr	57.3			
Cobalt	Co	64.6			
Copper	Cu	47.3			
Mercury	Hg	1.1			
Molybdenum	Mo	4.8			
Nickel	Ni	19.0			
Selenium	Se	6.1			
Silver	Ag	0.55			
Thallium	Tl	0.415			
Zinc	Zn	432.0			
ORGANIC COMPOUNDS			ORGANIC COMPOUNDS		
Trichloroethene	TCE	0.710	Trichloroethene	TCE	939
cis-1,2-Dichloroethene	cis-1,2-DCE	0.007	cis-1,2-Dichloroethene	cis-1,2-DCE	118

may both be exceeded by several orders of magnitude, as a result of receptor exposures to raw/untreated groundwater from the PLC site. The most significant contributors to site risks are from the general population exposure to manganese, thallium, TCE, and *cis*-1,2-DCE in groundwater.

A risk management decision Table 11.8 summarizes the results of the baseline risk assessment for the PLC site. In general, a cancer risk estimate greater than 10^{-6} or a noncarcinogenic hazard index greater than 1 indicates the presence of contamination that may pose a significant threat to human health. Overall, the levels of both carcinogenic and noncarcinogenic risks associated with the PLC site should *not* require time-critical removal action for contaminated soils present at this site. However, the use of untreated groundwater from aquifers underlying the site as a potable water supply source could pose significant risks to exposed individuals, or a community that uses such water for culinary purposes. Some type of corrective action or institutional control measure may therefore be necessary for this site.

The risk evaluation presented above indicates that the chemicals of potential concern present at the PLC site may pose some degree of risks to human receptors potentially exposed via the soil and groundwater media. Consequently, if this site is to be used for future commercial re-development projects, or if raw/untreated groundwater from this site is to be used as a potable water supply source, then a comprehensive site restoration program may be necessary to abate the imminent risks that the site poses especially from groundwater exposures. Whereas limited site control measures will probably be adequate to protect construction workers at the PLC site, a more extensive corrective action program will be needed for the impacted aquifer. Thus, general response actions (comprised of an integrated soil and groundwater remediation program) should be developed for each of the potentially impacted environmental media associated with the PLC site. Ultimately, the re-development of the site for commercial purposes may require only limited restoration activities, which may indeed be accomplished through 'incidental' site capping offered to the site following the construction of commercial buildings and pavements at the site.

11.4 A SITE RESTORATION STRATEGY

This section consists of a brief review of remedial action alternatives considered for the contaminated site problem described above. General response actions to consider and evaluate for the PLC site may include containment (e.g. via capping and immobilization techniques) and treatment (e.g. by soil vapor extraction, thermal treatments, biological treatments, and soil washing) for the soils; and pump-and-treat systems (including air or steam stripping, and granular activated carbon (GAC) adsorption) for groundwater. Table 11.9 presents a broader range of options considered and evaluated for this illustrative example. Ultimately, the remedial alternatives determined to be the most technically feasible become the focus for more detailed analyses. For other than a 'no-action' situation, the site-specific risk-based criteria developed above (shown in Tables 11.5 through 11.7) can be used to guide and support the development, screening, and selection of potentially feasible remedial alternatives.

146

Table 11.4 Tabular analysis chart for the exposure scenarios associated with the PLC site

Contaminated exposure medium	Contaminant release source(s)	Contaminant release mechanism(s)	Potential receptor location	Receptor groups potentially at risk	Potential exposure routes	Pathway potentially complete and significant?
Air	Contaminated surface soils	Fugitive dust generation	On-site	On-site facility worker	Inhalation Incidental ingestion Dermal absorption	Yes No No
				Construction worker	Inhalation Incidental ingestion Dermal absorption	Yes No No
			Off-site	Downwind worker	Inhalation Incidental ingestion Dermal absorption	No No No
				Downwind resident	Inhalation Incidental ingestion Dermal absorption	No No No
		Volatilization	On-site	On-site facility worker	Inhalation Dermal absorption	Yes No
				Construction workers	Inhalation Dermal absorption	Yes No
			Off-site	Nearest downwind worker	Inhalation Dermal absorption	No No
				Nearest downwind resident	Inhalation Dermal absorption	No No

Soils	Contaminated soils and/or buried wastes	Direct contacting	On-site	On-site facility worker	Incidental ingestion — Yes Dermal absorption — Yes
				Construction worker	Incidental ingestion — Yes Dermal absorption — Yes
			Off-site	Downwind worker	Incidental ingestion — No Dermal absorption — No
				Downwind resident	Incidental ingestion — No Dermal absorption — No
Surface water	Contaminated surface soils	Surface runoff into surface impoundments	On-site	On-site facility worker	Inhalation — No Incidental ingestion — No Dermal absorption — No
				Construction worker	Inhalation — No Incidental ingestion — No Dermal absorption — No
		Erosional runoff	Off-site	Downslope resident	Inhalation — No Incidental ingestion — No Dermal absorption — No
				Recreational population	Inhalation — No Incidental ingestion — No Dermal absorption — No
	Contaminated groundwater	Groundwater discharge	Off-site	Downslope resident	Inhalation — No Incidental ingestion — No Dermal absorption — No
				Recreational population	Inhalation — No Incidental ingestion — No Dermal absorption — No

(continued overleaf)

Table 11.4 (continued)

Contaminated exposure medium	Contaminant release source(s)	Contaminant release mechanism(s)	Potential receptor location	Receptor groups potentially at risk	Potential exposure routes	Pathway potentially complete and significant?
Groundwater	Contaminated soils	Infiltration/leaching	On-site	On-site facility worker	Inhalation	No
					Incidental ingestion	No
					Dermal absorption	No
				Construction worker	Inhalation	No
					Incidental ingestion	No
					Dermal absorption	No
			Off-site	Downgradient resident	Inhalation	Yes
					Incidental ingestion	Yes
					Dermal absorption	Yes
Drainage sediments	Contaminated surface soils	Surface runoff/ episodic overland flow	On-site	On-site facility worker	Inhalation	No
					Incidental ingestion	No
					Dermal absorption	No
				Construction worker	Inhalation	No
					Incidental ingestion	No
					Dermal absorption	No
			Off-site	Nearest downgradient resident	Inhalation	No
					Incidental ingestion	No
					Dermal absorption	No

149

Table 11.5 Risk screening for a nearby and/or on-site worker exposure to soils at the PLC site

Chemical of potential concern	Maximum soil concentration (mg/kg)	Chemical-specific dermal absorption (ABSs)	Oral RfD (mg/kg-day)	Oral SF (1/mg kg-day)	Inhalation RfD (mg/kg-day)	Inhalation SF (1/mg kg-day)	Risk for air	Hazard for air	Risk for soil	Hazard for soil	Total risk (air + soil)	Total hazard (air + soil)	Risk-based soil criteria (mg/kg)	Soil criteria exceeded?
INORGANIC CHEMICALS														
Antimony (Sb)	1.1	0.01	4.00E−04				0.00E+00	2.69E−05	0.00E+00	2.91E−03			38	No
Arsenic (As)	12.0	0.03	3.00E−04	1.75E+00		1.20E+01	5.03E−07	3.91E−04	1.64E−05	8.77E−02			0.71	Yes
Cadmium (Cd)	2.6	0.001	5.00E−04			1.50E+01	1.36E−07	5.09E−05	0.00E+00	2.84E−03			19	No
Chromium (Cr−total)	57.3	0.01	1.00E+00				0.00E+00	5.61E−07	0.00E+00	6.06E−05			937 615	No
Cobalt (Co)	64.6	0.01	2.90E−04		2.90E−04		0.00E+00	2.18E−03	0.00E+00	2.35E−01			272	No
Copper (Cu)	47.3	0.01	3.70E−02				0.00E+00	1.25E−05	0.00E+00	1.35E−03			34 692	No
Mercury (Hg)	1.1	0.01	3.00E−04		8.60E−05		0.00E+00	1.25E−04	0.00E+00	3.87E−03			275	No
Molybdenum (Mo)	4.8	0.01	5.00E−03		5.00E−03		0.00E+00	9.39E−06	0.00E+00	1.01E−03			4688	No
Nickel (Ni)	19.0	0.01	2.00E−02			9.10E−01	6.04E−08	9.30E−06	0.00E+00	1.00E−03			314	No
Selenium (Se)	6.1	0.01	5.00E−03				0.00E+00	1.19E−05	0.00E+00	1.29E−03			4688	No
Silver (Ag)	0.55	0.01	5.00E−03				0.00E+00	1.08E−06	0.00E+00	1.16E−04			4688	No
Thallium (Tl)	0.415	0.01	8.00E−05				0.00E+00	5.08E−05	0.00E+00	5.48E−03			75	No
Zinc (Zn)	432.0	0.01	3.00E−01				0.00E+00	1.41E−05	0.00E+00	1.52E−03			281 284	No
ORGANIC COMPOUNDS														
Trichloroethene (TCE)	0.710	0.10	6.00E−03	1.50E−02	6.00E−03	1.00E−02	2.48E−11	1.16E−06	2.34E−08	7.29E−04			30	No
cis-1,2-Di-chloroethene (DCE)	0.007	0.10	1.00E−02		1.00E−02		0.00E+00	6.85E−09	0.00E+00	4.32E−06			1620	No
							7.00E−07	0.003	1.65E−05	0.35	1.72E−05	0.35		

Notes
(1) The computational formulae and models used in this evaluation are discussed in Chapters 6 and 7.
(2) Risk/hazard for air accounts for only the airborne emissions of contaminated particulates for all chemicals present at the site; strict volatilization effects are not included in this screening analysis.
(3) Case-specific exposure parameters used in the calculations were obtained from the following sources – DTSC (1994), OSA (1993), and USEPA (1989a, 1989b, 1991, 1992).

Table 11.6 Risk screening for a construction worker exposure to soils at the PLC site

Chemical of potential concern	Maximum soil concentration (mg/kg)	Chemical-specific dermal absorption (ABSs)	Toxicity criteria				Risk for air	Hazard for air	Risk for soil	Hazard for soil	Total risk (air + soil)	Total hazard (air + soil)	Risk-based soil criteria (mg/kg)	Soil criteria exceeded
			Oral RfD (mg/kg-day)	Oral SF (1/mg kg-day)	Inhalation RfD (mg/kg-day)	Inhalation SF (1/mg kg-day)								
INORGANIC CHEMICALS														
Antimony (Sb)	1.1	0.01	4.00E-04				0.00E+00	5.38E-04	0.00E+00	1.45E-02			73	No
Arsenic (As)	12.0	0.03	3.00E-04	1.75E+00		1.20E+01	4.03E-07	7.83E-03	1.92E-06	2.56E-01			5	Yes
Cadmium (Cd)	2.6	0.001	5.00E-04			1.50E+01	1.09E-07	1.02E-03	0.00E+00	2.47E-02			24	No
Chromium (Cr)	57.3	0.01	1.00E+00				0.00E+00	1.12E-05	0.00E+00	3.02E-04			183 154	No
Cobalt (Co)	64.6	0.01	2.90E-04		2.90E-04		0.00E+00	4.36E-02	0.00E+00	1.17E-00			53	Yes
Copper (Cu)	47.3	0.01	3.70E-02				0.00E+00	2.50E-04	0.00E+00	6.73E-03			6 777	No
Mercury (Hg)	1.1	0.01	3.00E-04		8.60E-05		0.00E+00	2.50E-03	0.00E+00	1.93E-02			50	No
Molybdenum (Mo)	4.8	0.01	5.00E-03		5.00E-03		0.00E+00	1.88E-04	0.00E+00	5.05E-03			916	No
Nickel (Ni)	19.0	0.01	2.00E-02			9.10E-01	4.83E-08	1.86E-04	0.00E+00	5.00E-03			393	No
Selenium (Se)	6.1	0.01	5.00E-03				0.00E+00	2.39E-04	0.00E+00	6.42E-03			916	No
Silver (Ag)	0.55	0.01	5.00E-03				0.00E+00	2.15E-05	0.00E+00	5.79E-04			916	No
Thallium (Tl)	0.415	0.01	8.00E-05				0.00E+00	1.02E-03	0.00E+00	2.73E-02			15	No
Zinc (Zn)	432.0	0.01	3.00E-01				0.00E+00	2.82E-04	0.00E+00	7.58E-03			54 946	No
ORGANIC COMPOUNDS														
Trichloroethene (TCE)	0.710	0.10	6.00E-03	1.50E-02	6.00E-03	1.00E-02	1.98E-11	2.32E-05	1.58E-09	1.23E-03			444	No
cis-1,2-Di-chloroethene (DCE)	0.007	0.10	1.00E-02		1.00E-02		0.00E+00	1.37E-07	0.00E+00	7.26E-06			946	No
							5.60E-07	0.06	1.92E-06	1.55	2.48E-06	1.6		

Notes

(1) The computational formulae and models used in this evaluation are discussed in Chapters 6 and 7.

(2) Risk/hazard for air accounts for only the airborne emissions of contaminated particulates for all chemicals present at the site; strict volatilization effects are not included in this screening analysis.

(3) Case-specific parameters used in the calculations were obtained from the following sources – DTSC (1994), OSA (1993), and USEPA (1989a, 1989b, 1991, 1992).

Table 11.7 Risk screening for a downgradient residential population exposure to groundwater from the PLC site

Chemical of potential concern	Maximum water concentration (μg/L)	Chemical-specific Kp (cm/h)	Toxicity criteria				Risk for water	Hazard for water	Risk-based water criteria (μg/L)	Water criteria exceeded?
			Oral RfD (mg/kg-day)	Oral SF (1/mg/kg-day)	Inhalation RfD (mg/kg-day)	Inhalation SF (1/mg/kg-day)				
INORGANIC CHEMICALS										
Manganese (Mn)	1390	1.60E−04	5.00E−03		1.40E−05		0.00E+00	1.78E+01	78	Yes
Molybdenum (Mo)	6.5	1.60E−04	5.00E−03		5.00E−03		0.00E+00	8.31E−02	78	No
Thallium (Tl)	1.5	1.60E−04	8.00E−05				0.00E+00	1.20E+00	1.3	Yes
ORGANIC COMPOUNDS										
Trichloroethene (TCE)	939	1.60E−02	6.00E−03	1.50E−02	6.00E−03	1.00E−02	3.57E−04	2.02E+01	2.6	Yes
cis-1,2-Dichloroethene (DCE)	118	1.00E−02	1.00E−02		1.00E−02		0.00E+00	1.52E+00	78	Yes
							3.57E−04	40.7		

Notes
(1) The computational formulae and models used in this evaluation are discussed in Chapters 6 and 7.
(2) Risk/hazard for water accounts for both volatile chemical emissions and non-volatile chemical contributors present at the site (DTSC, 1994).
(3) Case-specific exposure parameters used in the calculations were obtained from the following sources – DTSC (1994), OSA (1993), and USEPA (1989a, 1989b, 1991, 1992).
(4) Kp = chemical-specific dermal permeability coefficient for water (DTSC, 1994; USEPA, 1992).

Table 11.8 Summary of the risk screening for the PLC site

Receptor group	Risk parameter	Exposure routes and pathways				
		Inhalation exposure to soils	Oral exposure to soils	Total exposure to soils	Total exposure to groundwater	Overall total risk
Hypothetical downgradient resident	Cancer risk	—	—	—	3.6×10^{-4}	3.6×10^{-4}
	Hazard index	—	—	—	40.7	40.7
Nearby and/or on-site worker	Cancer risk	7.0×10^{-7}	1.6×10^{-5}	1.7×10^{-5}	—	1.7×10^{-5}
	Hazard index	0.003	0.4	0.4	—	0.4
Construction worker	Cancer risk	5.6×10^{-7}	1.9×10^{-6}	2.5×10^{-6}	—	2.5×10^{-6}
	Hazard index	0.06	1.6	1.6	—	1.6

Note: Risk due to oral exposure is the total contribution from ingestion and dermal absorption of chemicals present in the contaminated medium.

Feasible remediation options An important site restoration technique for organic contamination in soils would be bioremediation; clearly identified 'hot spots' may however be removed for disposal or for treatment by other processes such as incineration. Other favored remediation techniques that appear more suitable for the PLC site would consist of the use of *in situ* soil venting systems. Inorganic contaminants that remain may be encapsulated or otherwise immobilized, or may be removed through soil washing or leaching activities.

The most frequently used treatment methods for volatile organic contamination in groundwater are air stripping and GAC adsorption. For greater removal efficiency, air stripping may be used in tandem with GAC (i.e. a GAC–air stripping remedial system). The overall remediation of the groundwater contamination at the PLC site may be accomplished more effectively by using a pump/treat/re-inject system. The preferred approach will be to pump from the edges of a mapped plume and re-inject (after treatment) at the center of the plume.

11.5 THE CORRECTIVE ACTION PLAN

The corrective action plan for the PLC facility may be divided into soil and groundwater remediation tasks. A soil vapor extraction (SVE) system is tentatively recommended for the impacted soils, and a pump-and-treat technology for the affected groundwater.

The soil remediation will consist of the design, installation, and operation of a SVE system (to remove residual VOCs from the soils, in order to prevent further impacts on groundwater), and laboratory analyses of verification samples (for monitoring purposes and performance evaluation).

The groundwater remediation will consist of the design, installation and operation of a groundwater treatment system (GTS), water sampling and analysis during operation, and laboratory analyses of verification samples (for monitoring purposes and performance evaluation).

The soil vapor extraction system The SVE system recommended for the PLC site restoration plan will utilize an array of vapor extraction wells and air injection wells located in areas of impacted soils, and designed to achieve maximum efficiency for contaminant removal; the use of the complementing air injection wells will generally improve the performance of the SVE system. Typically, upon implementation, the SVE system will be operated until the attainment of the acceptable cleanup levels for the key contaminants of concern in the contaminated soils. Subsequently, verification samples (to ascertain the performance of the SVE system) will be collected and analyzed prior to site closure or re-development.

The groundwater treatment system The GTS recommended for the PLC site restoration plan will be a pump-and-treat system that consists of both extraction and injection wells; the extraction–injection well network is designed and placed to achieve maximum efficiency for contaminant removal from the impacted groundwater. Groundwater pumped to the surface is treated prior to re-injection or disposal. Typically, upon implementation, the GTS will be operated until the attainment of the acceptable cleanup goals established for the key contaminants of concern in

Table 11.9 Screeening of a range of feasible remedial technologies and an evaluation of remediation process options for the PLC site

Target media for remediation program	General response action	Remedial technology	Preferred process option	Description of remediation program	Screening comments
Groundwater	No action	None	Not applicable	No action	Site remediation is required in order to achieve the remedial action objectives for this site
	Pump-and-treat	Groundwater extraction	Extraction/injection wells	Series of wells to extract contaminated groundwater; injection wells inject uncontaminated or treated water to increase flow to extraction wells	Preferred method of choice
	Containment	Capping (using low permeability caps and liners)	Clay cap	Compacted clay, covered with soil over areas of contamination	Potentially applicable as an effective alternative. Susceptible to cracking but has self-healing properties. Easily implemented. Low capital and maintenance costs
			Multimedia cap	Clay and synthetic membrane, covered with soil over areas of contamination	Potentially applicable as an effective alternative. Least susceptible to cracking. Easily implemented. Moderate capital and maintenance costs
		Vertical barriers	Slurry wall	Trench around area of contamination is filled with bentonite slurry	Potentially applicable
			Grout curtain	Pressure injection of grout in a regular pattern of drilled holes	Potentially applicable

	Vibrating beam	Vibrating force used to advance beams into the ground, with injection of slurry as beam is withdrawn	Potentially applicable
Horizontal barriers	Grout injection	Pressure injection of grout at depth through closely spaced drilled holes	Potentially applicable
	Block displacement	In conjunction with vertical barriers, injection of slurry in notched injection holes.	Potentially applicable
Gradient control	Hydraulic gradient manipulation	Use of hydraulic gradient to control flow	Potentially applicable
Bioremediation	Aerobic digestion	Degradation of chemicals using microorganisms in an aerobic environment	Potentially applicable
	Anaerobic digestion	Degradation of chemicals using microorganisms in an anaerobic environment	
Treatment	Physical treatment (e.g. physical separation or destruction) & chemical treatment (e.g. chemical modification or destruction)	Air stripping — Mixing large volumes of air and water in a packed column to promote transfer of VOCs to air	Potentially applicable to some contaminants

(continued overleaf)

Table 11.9 (continued)

Target media for remediation program	General response action	Remedial technology	Preferred process option	Description of remediation program	Screening comments
			Carbon adsorption	Adsorption of contaminants on to activated carbon by passing water through carbon column	Potentially applicable
			Reverse osmosis	Use of high pressure to force water through a membrane, leaving contaminants behind	Potentially applicable
			Incineration	Combustion	Potentially applicable
		Permeable treatment bed	Treatment bed	Treat shallow groundwater in-place by constructing permeable treatment beds that can physically and chemically remove contaminants	Potentially applicable as a temporary remedial measure
Site soils	No action	None	Not applicable	No action	Site remediation is required in order to achieve the remedial action objectives for this site
	Containment	Capping (using low permeability caps and liners)	Clay	Compacted clay, covered with soil over areas of contamination	Potentially applicable as an effective alternative. Susceptible to cracking but has self-healing properties. Easily implemented. Low capital and maintenance costs

	Multimedia cap	Clay and synthetic membrane, covered with soil over areas of contamination	Potentially applicable as an effective alternative. Least susceptible to cracking. Easily implemented. Moderate capital and maintenance costs
Vertical barriers	Slurry wall	Trench around area of contamination is filled with bentonite slurry	Potentially applicable
	Grout curtain	Pressure injection of grout in a regular pattern of drilled holes	Potentially applicable
	Vibrating beam	Vibrating force used to advance beams into the ground, with injection of slurry as beam is withdrawn	Potentially applicable
Horizontal barriers	Grout injection	Pressure injection of grout at depth through closely spaced drilled holes	Potentially applicable to overburden
	Block displacement	In conjunction with vertical barriers, injection of slurry in notched injection holes	Potentially applicable
Gradient control	Hydraulic gradient manipulation	Use of hydraulic gradient to control leachate flow	Potentially applicable
Excavation/ treatment	Excavation	Soils excavation/ removal	Mechanical removal of materials for treatment and/or re-disposal	Potentially applicable

(continued overleaf)

Table 11.9 (continued)

Target media for remediation program	General response action	Remedial technology	Preferred process option	Description of remediation program	Screening comments
		Immobilization/stabilization	Sorption and/or encapsulation	Use of pozzolanic agents to help immobilize site contaminants	Potentially applicable
		Physical treatment	Incineration and pyrolysis	Thermal treatment and methods	Potentially applicable, especially if NAPL should be encountered
		Chemical treatment	Neutralization	Use of chemical reagents to attenuate contaminant toxicity effects	Potentially applicable
		Biological treatment	Bioremediation	Use of microorganisms to degrade organic contaminants	Potentially applicable to some of the soil contaminants
	Vacuum extraction	Soil vapor extraction	Extraction/injection wells	Series of wells to extract contaminated soil vapors; injection wells inject fresh air to increase flow to extraction wells	Preferred method of choice

groundwater. Subsequently, verification samples (to ascertain the performance of the GTS) will be collected and analyzed prior to site closure; furthermore, long-term monitoring wells may be installed downgradient of the contaminated areas.

11.6 REFERENCES

DTSC (Department of Toxic Substances Control) (1994). *Preliminary Endangerment Assessment Guidance Manual* (A guidance manual for evaluating hazardous substance release sites). California Environmental Protection Agency, DTSC, Sacramento, California.

OSA (1993). *Supplemental Guidance for Human Health Multimedia Risk Assessments of Hazardous Waste Sites and Permitted Facilities*. Cal EPA, DTSC, 1993.

USEPA (US Environmental Protection Agency) (1989a). *Risk Assessment Guidance for Superfund*. Vol. 1 – *Human Health Evaluation Manual* (Part A). EPA/540/1-89/002. Office of Emergency and Remedial Response, Washington, DC.

USEPA (US Environmental Protection Agency) (1989b). *Exposure Factors Handbook*. EPA/600/8-89/043. Office of Health and Environmental Assessment, Washington, DC.

USEPA (US Environmental Protection Agency) (1991). *Risk Assessment Guidance for Superfund*. Vol. I – *Human Health Evaluation Manual*. Supplemental Guidance. *Standard Default Exposure Factors* (Interim Final). March 1991. Office of Emergency and Remedial Response, Washington, DC. OSWER Directive: 9285.6-03.

USEPA (US Environmental Protection Agency) (1992). *Dermal Exposure Assessment: Principles and Applications*. EPA/600/8-91/011B, January 1992, Washington, DC.

Chapter Twelve

Design of corrective action response programs

The design of corrective action response programs for contaminated site problems usually involves several formalized steps. Typically, the major activities carried out will comprise of a remedial investigation (RI) and a feasibility study (FS). The RI is conducted to gather sufficient data in order to characterize conditions at a contaminated site; this may involve extensive fieldwork and site assessment used to identify site contaminants, determine constituent concentrations and distribution across the site, characterize contaminant migration pathways and routes of exposure to local populations, and identify other site conditions that might affect site restoration options. The FS is undertaken to identify and analyze potential site restoration alternatives; typically, the FS uses a screening process to reduce the number of alternative corrective measures to a limited range of remedial options, and the short-listed options are subsequently subjected to a detailed analysis in which tradeoffs (such as cost-effectiveness, extent of cleanup, and permanence of a cleanup action) are evaluated.

A variety of corrective action strategies may indeed be employed in the quest to restore contaminated sites into healthier conditions. In the process involved, many complex and interacting factors tend to influence the development of the site restoration program. In particular, a clear understanding of the fate and behavior of the contaminants in the environment is essential for developing successful corrective action response programs, and also to ensure that the problem is not exacerbated.

This chapter presents strategies and concepts that can generally be used to aid contaminated site management decisions. In general, once existing site information has been analyzed and a conceptual understanding of a site is obtained, potential remedial action objectives should be defined for all impacted media at the contaminated site. Subsequently, alternative site restoration programs can be developed to support the requisite corrective action decision.

12.1 GENERAL SCOPE OF THE RI/FS PROCESS

Typically, the following general tasks are carried out as part of the overall RI/FS program designed to address contaminated site problems (Cairney, 1993; OBG, 1988; USEPA, 1988):

- Characterize site as completely as possible.
- Define and develop remediation objectives that are appropriate for the specific contamination problems.

- Identify remedial technologies that are capable of achieving the site restoration goal(s).
- Develop and screen remedial alternatives, and select only those that are superior based on engineering, environmental, and economic criteria, and which concurrently meet regulatory requirements and community expectations.
- Perform detailed analyses of remedial alternatives by scrutinizing each of the initially selected alternatives.
- Choose, offering justification, the preferred remedial alternative(s).

Information derived from the RI will generally help determine the need for, and the extent of cleanup necessary for a contaminated site. The early determination of remediation goals facilitates the development of a range of feasible site restoration plans, which in turn helps focus remedy selection on the most effective remedial alternative(s). Upon selection of one of a number of cleanup options identified in the FS, a project then enters an engineering design phase in which plans and specification are developed for the selected remedy. Once the engineering design is completed, construction activities to deal with the actual cleanup of a site can be implemented. In some cases, such as is atypical of groundwater remediation, cleanup activities may continue for several years.

12.2 SITE ASSESSMENT AND CHARACTERIZATION CONSIDERATIONS

Site assessments are, invariably, a primary activity in the overall process involved in the management of contaminated site problems. The objective of the site assessment is to determine the nature and extent of potential impacts from the release or threat of release of hazardous substances.

A site investigation effort, which is a major component of the site assessment activity, aims at collecting representative samples from the potentially contaminated site. Depending on the adequacy of historical data and the sufficiency of details about the likely contaminants at a site, sampling programs can be designed to search for specific chemical constituents that become indicator parameters for the sample analyses. Where specific information about historical uses of the site is lacking, a more comprehensive sampling and analytical program will generally be required; in this case, the sampling and analysis program may be carried out in phases – moving from a more general scope to one of specificity as adequate information becomes available about the site. The results from these activities will facilitate a complete analysis of the possible contaminants present at the site. The information obtained is used to determine current and potential future risks to human health and the environment. On this basis, corrective actions are developed and implemented, with the principal objective to protect public health and the environment.

12.2.1 The Site Characterization

When there is a source release, contaminants may be transported to different potential receptors via several environmental media (such as air, soils, groundwater and surface water). A complexity of processes may affect contaminant migration at contaminated sites, resulting in human and ecological receptors outside the source area potentially being threatened. Consequently, it is imperative to adequately characterize a site and its surroundings through a well-designed site assessment program (in which all contaminant sources and impacted media are thoroughly investigated), in order to arrive at appropriate and cost-effective corrective action response decisions.

A multimedia approach to site characterization is usually adopted for most contaminated site problems, so that the significance of possible air, water, soil, and biota contamination can be established through appropriate field sampling and analysis procedures. The activities involved are expected to yield high-quality environmental data needed to support the corrective action response decision. To accomplish this, samples are gathered and analyzed for the contaminants of potential concern in the appropriate media of interest. Proper protocols in the field sampling and laboratory analysis procedures are used to minimize uncertainties associated with the data collection and evaluation activities.

Ultimately, the information gathered during the site investigation activities are used to map out the extent of contamination, and to evaluate the potential risks associated with the subject site so that an appropriate corrective action response decision can be made. Invariably, health and environmental risk estimates become an important element in the corrective action decision process. This is because the risk assessment provides the decision-maker with technically defensible and scientifically valid procedures for determining whether or not a contaminated site represents significant adverse human health or ecological risks that warrant consideration as a candidate for mitigation.

12.3   THE ROLE OF RISK ASSESSMENT

The assessment of health and environmental risks plays an important role in site characterization activities, in corrective action planning, and also in risk mitigation and risk management strategies for contaminated site problems. A major objective of any contaminated site risk assessment is to provide an estimate of the baseline risks posed by the existing conditions at the site, and to further assist in the evaluation of site restoration options.

The application of the risk assessment process to contaminated site problems will generally serve to document the fact that risks to human health and the environment have been evaluated and incorporated into the appropriate response actions. In fact, it is apparent that some form of risk assessment is inevitable if site characterization and corrective action response programs are to be conducted in a sensible and deliberate manner. This is because, the very process of performing risk assessment does lead to a better understanding and appreciation of the nature of the risks inherent in a study, and further helps develop steps that can be taken to reduce such risks. Appropriately applied, risk assessment techniques can be used to estimate the risks posed by site contaminants under various exposure scenarios, and to further estimate the degree of risk reduction achievable by implementing various engineering remedies. Almost invariably, every process for developing corrective action response strategies should incorporate some concepts or principles of risk assessment. In particular, all decisions on site restoration plans for contaminated sites will include, implicitly or explicitly, some elements of risk assessment.

Overall, the application of risk assessment to contaminated site problems helps identify critical migration and exposure pathways, receptor exposure routes, and other extraneous factors contributing most to total risks. It also facilitates the determination of cost-effective risk reduction policies. Used in the corrective action planning process, risk assessment generally serves as a useful tool for evaluating the effectiveness of remedies at contaminated sites, and also for determining acceptable cleanup levels. Inevitably, risk-based corrective action programs facilitate the selection of appropriate and cost-effective site restoration measures.

12.3.1 The Acceptable Cleanup Criteria

Risk assessment is particularly useful in determining the level of cleanup most appropriate for contaminated sites. By utilizing methodologies that establish cleanup criteria based on risk assessment principles, site restoration programs can be carried out in a cost-effective and efficient manner.

To arrive at a responsible decision on what acceptable cleanup criteria to adopt for a contaminated site problem, a site restoration program should, among other things, carefully evaluate the following important parameters:

- The *level of risk* indicated by the contaminants of concern.
- The *background threshold concentration levels* for the contaminants of concern at upgradient, upstream, and/or upwind locations relative to the source(s) of contamination or release(s).
- The *natural attenuation effects* of the contaminants of concern. Attenuation may result from several processes such as evaporation, photolysis, dilution, biodegradation, etc.
- The *asymptotic level* of the contaminants of concern in the impacted media, which represents the cleanup level corresponding to a point of diminishing returns (i.e. the point when little additional progress can be made in reducing the contaminant levels). This corresponds to the attainment of contaminant levels below which continued remediation produces negligible reductions in contaminant levels.
- The *best demonstrated available technologies* that can be proven to offer feasible and cost-effective remediation techniques to the site restoration program.

Other principal considerations about the site restoration goal relate to the cost of cleanup, time required to complete site remediation, and the possibility of a cleanup activity creating potential liability problems. In fact, once realistic risk reduction levels potentially achievable by various remedial alternatives are known, the decision-maker can use other scientific criteria (such as implementability, reliability, operability, and cost) to select a final design alternative. Subsequently, an appropriate corrective action plan can then be developed and implemented for the contaminated site.

12.4 SITE RESTORATION CONSIDERATIONS

The site restoration goal and strategy selected for a contaminated site problem may vary significantly from one site to another due to the potential effects of several site-specific parameters. A number of extraneous factors may also affect the selection of site restoration strategies. Ultimately, however, remedial action alternatives selected for contaminated site problems should be designed and operated in such a manner as to be fully protective of human health and the environment.

In general, the selection of an appropriate corrective action response plan depends on a careful assessment of both short- and long-term risks posed by the case-specific site. As a general rule, corrective action response plans should provide for the removal and/ or treatment of contaminants until a level necessary to protect human health and the environment is achieved.

12.4.1 Site Categorization

A broad categorization scheme for a variety of potentially contaminated area designations may be used to facilitate the planning, development and implementation of appropriate site restoration or closure programs. Sites may indeed be clustered into different groups in accordance with the level of effort required to implement the appropriate or necessary response actions. Typical designations, which are by no means exhaustive, are enumerated below.

- *Low-risk areas*, represented by sites with no confirmed contamination, may consist of areas (e.g. suspected sources) where the results of records search and site investigations show that no hazardous substances were stored for any substantial period of time, released into the environment or site structures, or disposed of on the site property; or areas where the occurrence of such storage, release, or disposal is not considered to have been probable. The determination of a low-risk area can be made at any of several decision stages during the site assessment.
- *Intermediate-risk areas*, represented by sites with limited contamination, may consist of impacted areas where no response or remedial action is necessarily required to ensure protection of human health and the environment. Intermediate-risk areas include areas where an environmental assessment may have demonstrated that hazardous materials have been released, stored, or disposed of, but are present in quantities that probably require only limited to no response or remedial action to protect human health and the environment. Such designation means that the levels of hazardous substances detected in a given area *do not* exceed media-specific action levels (e.g. risk-based concentrations); *do not* result in significant risks; *nor* otherwise exceed requisite regulatory standards.
- *High-risk areas*, represented by sites with extensive contamination, may consist of areas where the site records indicate that hazardous materials are known to have been released or disposed of. Typically, a significant level of corrective action response will be required for such sites.

Depending on the site category, different site restoration strategies may be utilized in the corrective action response program.

12.4.2 Interim Corrective Action Programs

Interim corrective actions are measures used to address situations which pose imminent threat to human health or the environment, or to prevent further environmental degradation or contaminant migration, pending final decisions on the necessary long-term remedial activities.

Whenever excessive risks exist at a contaminated site, a decision should be made to implement interim corrective actions immediately, in order to protect public health and the environment. Common examples of interim corrective measures may include simply erecting a fence around a contaminated site in order to restrict/limit access to an impacted property; covering exposed contaminated soils with synthetic liner materials; applying dust suppressants to minimize emissions of contaminated fugitive dust; restricting use of contaminated groundwater, and/or providing alternative sources for culinary water supplies; temporal displacement or relocation of nearby residents away from a hazardous waste site; and the implementation of more elaborate control measures such as installing a pump-and-treat system to prevent further migration of a groundwater contaminant plume.

12.4.3 Documentation of 'No-further-action' Decisions

A major reason for conducting site investigation and characterization activities at contaminated sites is to be able to make informed decisions regarding site restoration programs; the fundamental purpose of a site restoration program is to protect human health and the environment from the unintended consequences of environmental contamination problems. In some situations, a decision that 'no further action' (NFA) is required for a site may be appropriate for a contaminated site problem. The process involved in a NFA decision (NFAD) is intended to indicate that, based on the best available information, no further response action is necessary to ensure that the case site does not pose significant risks to human health or the environment. Under such circumstances, a NFA closure document will usually be prepared for the site or group of sites, in accordance with applicable regulatory requirements. The NFA document usually is a stand-alone report, containing sufficient information to support the NFAD. It should therefore include site-specific evidence along with adequate technical reasoning and justification for the NFAD. Additionally, the NFA document should clearly address specific CSM hypotheses that have been tested to confirm that there are no likely and complete exposure pathways or scenarios associated with the site.

Contaminant cleanup standards specified by several regulations (such as media-specific action levels, site-specific risk-based criteria, and local or area background threshold levels) often form an important basis for a NFAD. Ultimately, however, the NFA determination rests on whether or not complete exposure pathways exist at a site, and whether or not any of the complete pathways is significant. It is therefore logical to infer that the NFA decision criteria are indeed linked to the use of the CSM as a decision tool. Consequently, an evaluation of the elements or components of a CSM will generally serve as an additional important basis for many NFADs. This typically involves a demonstration that the source–pathway–receptor linkage cannot be completed at the site, or that if the linkage can be completed the risk posed by the contamination present does not exceed 'acceptable' reference standards established for the site or area.

12.4.4 Ecological Risk Considerations in Corrective Action Decisions

Oftentimes, and more so in the past, only limited attention has been given to the ecosystems associated with contaminated sites, and also to the protection of ecological resources during site remediation activities. Instead, much of the focus has been on the protection of human health and resources directly affecting public health and safety. In recent times, however, the ecological assessment of contaminated sites is gaining considerable attention. This is the result of prevailing knowledge or awareness of the intricate interactions between ecological receptors/systems and contaminated site cleanup processes.

In a number of situations, site remediation can destroy or otherwise affect uncontaminated ecological resources; soil removal techniques, alteration of site hydrology, and site preparation are examples of remediation activities that can result in inadvertent damage to ecological resources. Thus, the ecological impacts of a site restoration activity must be understood by the decision-makers before remediation plans are approved and implemented. In situations where adverse ecological effects are identified, corrective action alternatives with potentially less damaging impacts must be evaluated as preferred methods of choice. This means that the assessment of potential ecological impacts should be performed for each remedial alternative considered for a contaminated site.

To achieve adequate ecological protection and/or regulatory compliance, ecological assessments should address the overall site contamination issues and should be coordinated with all aspects of site cleanup, including human health concerns, engineering feasibility, and economic considerations (Maughan, 1993). In fact, there are several important ecological concerns associated with contaminated site cleanup programs that should be addressed early enough in site characterization programs. Furthermore, a number of legislative requirements for incorporating ecological issues in site characterization and site restoration efforts now exist in several geographical regions, and these can no longer be ignored. It is apparent that ecological resources must be appropriately evaluated in order to achieve the mandate of any comprehensive program designed to ensure the effective management of contaminated site problems.

12.5   CORRECTIVE ACTION ASSESSMENT AND RESPONSE PROGRAMS

Corrective action assessment programs are generally designed to facilitate the development of feasible scientifically based methods necessary to support corrective action response decisions. The typical tasks performed in these types of programs may be comprised of: a site assessment; a risk analysis; an evaluation of remedial alternatives (in terms of cost, probable effectiveness, etc.); the ranking of feasible remedial options; a recommendation of the overall best remedial technology; and the design and implementation of requisite remedial options, together with monitoring programs (Figure 12.1). The process allows for a multimedia approach to site characterization; identifies the specific parameters for which data must be collected; identifies the preferred data gathering, handling, and analysis techniques to be used; and allows project managers to develop site-specific and cost-effective cleanup solutions. In general, once quality-assured information has been compiled for a potentially contaminated site and a benchmark risk level is established, an acceptable cleanup criterion can be determined that will be used to guide further site restoration decisions.

Corrective action response programs for contaminated sites may indeed vary greatly, ranging from a 'no-action' alternative to a variety of extensive and costly remediation

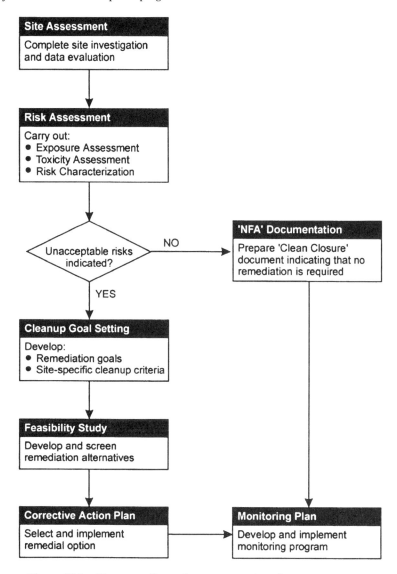

Figure 12.1   The corrective action assessment and response program

or site restoration options, such as those discussed in Chapter 8. In any case, the primary objective of every corrective action response program is to ensure public safety and welfare by protecting human health, the environment, and public and private properties. The ability to select an appropriate and cost-effective site restoration strategy that meets this goal will generally depend on a careful assessment of both short- and long-term risks potentially posed by the site (determined by procedures described in Chapter 6); it also depends on the site-specific restoration goal that is established and accepted for the site (determined by methods presented in Chapter 7), as well as the discretionary use of several corrective measure evaluation tools such as those that were alluded to in Chapter 9.

In all cases of contaminated site studies, an initial qualitative evaluation of the site conditions will help identify potential contaminant release source(s), determine the environmental media affected by each release, and broadly define the possible extent of the release(s). For a confirmed release, the release source(s) must be stopped and hazards mitigated. Where applicable, any free product present should be removed immediately to prevent the development of health and safety hazards; removal will also help prevent further migration of free product into soils and/or groundwater. The questionnaire chart included in Appendix E can be used to facilitate this process.

12.5.1 Decision Elements Associated with the Design of Corrective Action Programs

Corrective actions are generally designed with the goal to minimize potential negative impacts associated with contaminated site problems. The processes involved in the design of an adequate strategy will normally incorporate a consideration of the complex interactions existing between the hydrogeological environment, regulatory policies, and the technical feasibility of remedial technologies. Typically, several pertinent questions relating to the nature and extent of contamination, exposure settings, migration and exposure pathways, populations potentially at risk, the nature and level of risks, and cleanup goals are asked during the planning, development and implementation of corrective action programs that are directed at restoring contaminated sites (BSI, 1988; Cairney, 1993; Jolley and Wang, 1993; USEPA, 1985, 1987a, 1987b, 1988, 1989, 1991; WPCF, 1988). It is very important that the corrective action assessment helps answer all relevant questions, which will ultimately affect the type of corrective action response decision accepted for a contaminated site problem. In particular, the design of an effectual corrective action program for any contaminated site problem should account for the following important decision elements:

- Characterization of the contaminated site, including the physical setting, site geology, topography, hydrogeology and meteorological conditions.
- Identification of contaminant types and their characteristics.
- Assessment of the fate and transport characteristics of site contaminants, including an identification of anticipated degradation, reaction and/or decomposition byproducts.
- Determination of the critical environmental media of concern (such as air, surface water, groundwater, soils and sediments, and terrestrial and aquatic biota).
- Delineation of potential migration pathways.
- Identification and characterization of potential human and ecological receptors.
- Development of a conceptual representation or model of the site.
- Evaluation of potential exposure scenarios, and the possibility for human and ecosystem exposures.
- Assessment of the environmental and health impacts of site contaminants, if they should reach critical human and ecological receptors.
- Determination of the corrective action needs, and development of a risk management strategy for the site.
- Design of effective long-term monitoring and surveillance programs as a necessary part of an overall corrective action plan.

These basic elements should be completely evaluated and the relevant information used to support the appropriate corrective action decisions. Although different levels of effort may be required for different sites, a comprehensive site assessment program is always essential for making good corrective action decisions. In fact, an adequately conducted site assessment will allow a project manager to develop and select a final corrective action plan or cleanup solution that best suits the conditions at a particular site. The nature of this assessment will generally depend on the stage of investigation.

12.5.2   A Framework for Corrective Action Programs

The key components of a typical corrective action assessment and response program will be comprised of the following tasks:

- A preliminary site appraisal
- A site assessment
- A risk appraisal
- A risk determination
- A site mitigation.

The pertinent ingredients of these tasks are presented below.

Information derived from the relevant evaluations will generally help formulate credible policies to support the corrective action program. In general, the use of a systematic evaluation process will particularly facilitate rational decision-making on site restoration efforts that are undertaken for contaminated sites.

The preliminary site appraisal   The purpose of a *preliminary site appraisal*, consisting of the identification of possible source(s) of contaminant releases, is to quickly assess the potential for a site to adversely impact the environment and/or public health. This process typically involves: establishing a basis for contamination indicators; gathering and reviewing site background information for evidence of release; determining the site history, to help identify other possible sources of contamination; identifying potentially affected areas; addressing health and safety issues associated with the case-specific situation; and addressing emergency response by mitigating release and potential hazards. The site appraisal is initiated by the discovery of a potentially contaminated site. Conventional site reconnaissance procedures may be used in this qualitative site assessment that involves the collection and review of all available information (including an off-site reconnaissance to evaluate the source and nature of contamination, and the identification of any potential 'outside' polluters). Depending on the results of the preliminary survey, a site may or may not be referred for further action. In general, the preliminary site appraisal allows for site screening that will help establish a basis for more detailed site investigations.

The site assessment   The objectives of a *site assessment*, involving the characterization of site contamination and a site categorization (where necessary), are to identify site contaminants and to determine site-specific characteristics that influence the migration of the contaminants. The process typically involves the following: characterize site (to include an identification of the contaminants of potential concern and a delineation of the extent of contamination for affected matrices); and determine contaminant behavior in the environment, by developing a working hypothesis about contaminant fate and transport. Overall, site assessment activities are undertaken to define more completely the characteristics of a potentially contaminated site and its neighboring areas.

The risk appraisal The primary objective of the *risk appraisal*, consisting of a determination of the migration and exposure pathways integrated with an assessment of the environmental fate and transport of the contaminants of concern, is to determine whether potential receptors are likely to be at risk as a result of exposure to site contaminants. The process typically involves the following: determine contaminant migration and exposure pathways; identify populations potentially at risk; map areas where contaminants may impact human health and/or environment; develop realistic exposure scenarios appropriate for the specific site; and conduct site-specific exposure assessments. To determine if potential receptors are at risk, it is necessary to identify potential migration and exposure pathways as well as contaminant exposure point concentrations in relation to acceptable threshold levels.

The risk determination The objective of the *risk determination*, to include an evaluation of the environmental and health impacts associated with contaminant releases as well as the development of site-specific cleanup criteria, is to evaluate potential site risks. The calculated risk can then be compared against a benchmark risk, which then forms the basis for developing site restoration goals. The process typically involves defining the magnitude or severity of health and environmental impacts.

The site mitigation The *site mitigation* consists of the development and implementation of a corrective action plan that may include site remediation and/or monitoring programs. The process typically involves the following: establish remediation objectives and site-specific cleanup criteria; assess variables influencing selection of cleanup criteria and remedial systems; develop remedial action objectives; identify remedial action alternatives; and develop general response actions. Site mitigation strategies and objectives are based on the protection of both current and potential future receptors that could become exposed to site contaminants. Consequently, the site restoration is carried out so as to leave the site in such a condition as to pose no significant risks to any populations that may be potentially at risk.

12.5.3   The Need for Monitoring Programs

In addition to adopting appropriate remedial actions, monitoring programs will normally be implemented to verify the long-term effectiveness of an overall corrective action program. Monitoring is indeed considered a very important component of corrective action programs, and can serve as a useful tool for evaluating the performance of site restoration actions. In this regard, several different types of monitoring programs may be employed to evaluate the effectiveness of a remedy. For instance, an environmental monitoring program is normally used to ensure that chemical loadings from contaminated sites do not continually escape the influence of applicable remedial systems. Where necessary, the remedial system is modified to ensure that the site restoration program is completely effective – within the limits of any technical constraints.

Several monitoring parameters are important to the design of an effectual monitoring program. Of special interest is the selection of monitoring constituents. In general, the selection of monitoring constituents should consider the possibility for chemical compounds to be transformed over time and space. For instance, knowledge about the degradation of contaminants can be an extremely important factor in identifying monitoring constituents. This is because physical, chemical, and biological degradation may transform certain constituents as the release ages or advances. Thus, specific monitoring constituents and indicator parameters may have to be modified as an investigation progresses in time, in order to account for transformation products. In fact, despite the notion that most chemicals usually degrade into less toxic, more stable species, this is not universally true; for example, one of the degradation products of trichloroethene (TCE) is vinyl chloride, and both of these are carcinogenic chemicals. Consequently, the potential for physical, chemical, or biological transformations of constituents should be given adequate consideration in identifying monitoring constituents to be used in a corrective action program.

12.6 REMEDIATION OF CONTAMINATED SOILS

Several remedial options exist for the handling of soil contamination problems. Prominent among these are the following broad categories (Charbeneau et al., 1992; Sims et al., 1986): containment (e.g. encapsulation), source removal (including source excavation), *in situ* treatment, and vacuum extraction. Depending on the site-specific situation, it often becomes necessary to employ more than one technique in order to accomplish a soil remediation task. The utility of any one soil remediation technique invariably depends on the nature of contamination and the level of risks posed by the contaminants. In any case, an in-place remedial technology or process will generally be given preference as a method of choice in the determination of soil remediation options for a contaminated site problem. In-place remedial techniques may include extraction (e.g. soil washing); immobilization (e.g. sorption, ion exchange, precipitation); chemical degradation (e.g. degradation involving oxidation, reduction, and polymerization reactions); biodegradation (using natural microorganisms or genetically engineered microbes); photolysis (which may include enhanced photodegradation achieved by the addition of proton donors in the form of polar solvents); attenuation (e.g. mixing of contaminated surface soil with clean soil); and reduction of votalization effects (Sims et al., 1986). Typically, the in-place technique is used to contain the source of contamination, or to remove contamination through treatment processes.

A primary consideration in the identification of appropriate site-specific in-place technologies for contaminated soils relates to the chemical nature of the contaminants of concern. In fact, the design and implementation of an in-place treatment process requires information on characteristics of the contaminant/soil systems as a whole, with particular attention given to the following important variables (Charbeneau et al., 1992; Sims et al., 1986; USEPA, 1984):

- *Depth to contamination*. If contamination is limited to the upper 15–20 centimeters ($\approx$ 6–8 inches) of the soil and is well above the water table, in-place treatment techniques may be much more easily applied than if the contamination extends well below the ground surface and into a seasonally high water table.
- *Contaminant concentrations and quantities*. The efficiency and effectiveness of an in-place process depends on both contaminant concentration levels and quantity of each contaminant present in a given area.

In general, contaminant and soil characteristics are used to pre-screen in-place alternatives for their potential applicability in meeting site restoration goals. Based on the contaminant, soil, and system characteristics, an analysis can be made of the pathways and rates of contaminant migration, and also the potential for damage to human health and the environment as a function of time under 'no action' conditions.

The overall details of commonly used soil restoration techniques used to address site contamination problems were previously discussed in Chapter 8.

12.6.1   Treatability and Bench-scale Investigations

If remedial actions involving treatment are identified for a contaminated site problem, and if existing site and/or treatment data are insufficient to evaluate adequately such alternatives, then the need for treatability or bench-scale studies should be determined as early as possible in the corrective action assessment process.

Treatability and bench-scale tests may be necessary to evaluate the effects of a particular technology on specific site contaminants. Such tests will typically involve pilot testing to gather information that will help assess the feasibility of selected technologies. Treatability studies are conducted to provide additional data required for the evaluation of relevant soil treatment alternatives, and to support the remedial design of a selected alternative (USEPA, 1988). In some situations, a pilot-scale study may be necessary to furnish performance data and to develop better cost estimates so that a detailed analysis can be carried out to aid the selection of a remedial option.

Ultimately, it will help reduce cost and performance uncertainties for treatment alternatives to acceptable levels, resulting in the implementation of an appropriate site restoration program.

12.7   REMEDIATION OF GROUNDWATER CONTAMINATION PROBLEMS

Several remedial alternatives exist for the handling of groundwater contamination problems. Prominent among these are the following broad categories (Charbeneau et al., 1992): containment, source removal (including pumped removal of product and/or contaminated water), *in situ* treatment (chemical or biological), and vacuum extraction.

Depending on the site-specific situation, it often becomes necessary to use more than one method in order to complete remediation in the saturated and vadose zones.

Methods for mitigating contaminated groundwater plumes typically include the use of physical containment systems (such as impermeable barriers, hydraulic barriers, and subsurface collection systems) and leachate controls.

Containment systems act to interrupt contaminant transport mechanisms in order to prevent or minimize the continuing spread of contamination. Impermeable barriers may consist of slurry walls, grout curtains, and sheet piling; these can be used to contain, capture or redirect groundwater flow for the mitigation of groundwater contamination problems. Hydraulic barriers (e.g. recovery wells, interceptor trenches, etc.) are used to modify hydraulic gradients around contained waters; these can be used to manipulate, through pumping/injection strategies, the movement and size of a contaminant plume – given the proper subsurface conditions.

Leachate controls may include capping (to prevent or minimize rainwater from infiltrating through contaminated soil into groundwater) and/or the use of subsurface drains (consisting of buried conduits that collect and convey leachate).

Groundwater pump-and-treat systems, consisting of extraction/injection well networks have been a very commonly used remediation technique. In the remedial design for groundwater contamination problems, the recovery well systems are designed to intercept the contaminant plume so that no further degradation of the impacted aquifer occurs. Typically, groundwater treatment technologies applicable to the recovered groundwaters include physical treatment processes (such as phase-separated hydrocarbon recovery, air stripping, activated carbon adsorption, and filtration – which are all processes generally applied without the aid of chemical or biological agents); chemical processes (such as coagulation–precipitation, oxidation–reduction [redox], neutralization, etc.); and biological methods (such as suspended-growth and fixed-film reactors, as well as *in situ* biodegradation).

General details of common technologies and processes employed for the prevention and/or cleanup of groundwater contamination were previously discussed in Chapter 8.

12.7.1 The Challenges of Contaminated Aquifer Restoration Programs

The restoration of contaminated aquifer systems is one of the most challenging problems in corrective action response programs. Oftentimes, far too much attention is given to pump-and-treat remedial technologies. However, this technique may leave a great part of contaminant residues in the capillary fringe or vadose zone, unaffected by groundwater pumping. Removal of the contaminated soil at the case site, followed by treatment and/or disposal, will generally be a better strategy to adopt in order to address source elimination. In fact, in most aquifer contamination problems, containment of the aquifer contaminants is the most immediate concern. This can be achieved through the use of a physical barrier (such as a grout curtain, slurry cut-off wall or sheet piles), or by the creation of a hydraulic barrier resulting from a network of extraction (pumping) and injection (recharge) wells. The next requirement will be the removal of the mobile contaminants from the aquifer system. This may include free product recovery (e.g. of petroleum products floating on a water table), air stripping and vacuum extraction of the volatile organic contaminants, bioremediation, and a pump-and-treat technology for the soluble constituents. Insoluble constituents may remain adsorbed on to soil particles in the aquifer or vadose zone, to be addressed differently.

12.8 REFERENCES

BSI (British Standards Institution) (1988). Draft for Development, DD175: 1988 Code of Practice for the Identification of Potentially Contaminated Land and its Investigation. BSI, London, UK.

Cairney, T. (ed.) (1993). *Contaminated Land (Problems and Solutions)*. Blackie Academic & Professional, Glasgow/Chapman & Hall, London/Lewis Publishers, Boca Raton, Florida.

Charbeneau R.J., P.B. Bedient, and R.C. Loehr (eds) (1992). *Groundwater Remediation. Water Quality Management Library*, Vol. 8. Technomic Publishing Co., Inc., Lancaster, PA.

Jolley, R.L. and R.G.M. Wang (eds) (1993). *Effective and Safe Waste Management: Interfacing Sciences and Engineering With Monitoring and Risk Analysis*. Lewis Publishers, Boca Raton, Florida.

Maughan, J.T. (1993). *Ecological Assessment of Hazardous Waste Sites*. Van Nostrand Reinhold, New York.

OBG (O'Brien & Gere Engineers, Inc.) (1988). *Hazardous Waste Site Remediation: The Engineer's Perspective*. Van Nostrand Reinhold, New York.

Sims, R. et al. (1986). *Contaminated Surface Soils In-Place Treatment Techniques*. Noyes Publications, Park Ridge, New Jersey.

USEPA (US Environmental Protection Agency) (1984). *Review of In-place Treatment Techniques for Contaminated Surface Soils*, Vols 1 and 2. US Environmental Protection Agency, Hazardous Waste Engineering Research Laboratory, Cincinnati, OH. EPA-540/2-84-003a and b.

USEPA (US Environmental Protection Agency) (1985). *Characterization of Hazardous Waste Sites: A Methods Manual*, Vol. 1 – *Site Investigations*. US Environmental Protection Agency, Environmental Monitoring Systems Laboratory, Las Vegas. EPA-600/4-84-075.

USEPA (US Environmental Protection Agency) (1987a). *Alternate Concentration Limit Guidance*. Report No. EPA/530-SW-87-017, OSWER Directive 9481-00-6C, USEPA, Office of Solid Waste, Waste Management Division, Washington, DC.

USEPA (US Environmental Protection Agency) (1987b). *RCRA Facility Investigation (RFI) Guidance*. EPA/530/SW-87/001, Washington, DC.

USEPA (US Environmental Protection Agency) (1988). *Guidance for Conducting Remedial Investigations and Feasibility Studies Under CERCLA*. EPA/540/G-89/004. OSWER Directive 9355.3-01, Office of Emergency and Remedial Response, Washington, DC.

USEPA (US Environmental Protection Agency) (1989). *Risk Assessment Guidance for Superfund*. Vol. I – *Human Health Evaluation Manual* (Part A). EPA/540/1-89/002. Office of Emergency and Remedial Response, Washington, DC.

USEPA (US Environmental Protection Agency) (1991). *Risk Assessment Guidance for Superfund*. Vol. I – *Human Health Evaluation Manual*. Supplemental Guidance. *Standard Default Exposure Factors* (Interim Final). March 1991. Office of Emergency and Remedial Response, Washington, DC. OSWER Directive: 9285.6-03.

WPCF (Water Pollution Control Federation) (1988). *Hazardous Waste Site Remediation: Assessment and Characterization*. A Special Publication of the WPCF, Technical Practice Committee, Alexandria, Virginia.

Additional bibliography

Acar, Y.B. and D.E. Daniel (eds) (1995). *Geoenvironment 2000: Characterization, Containment, Remediation, and Performance in Environmental Geotechnics, Vols 1 and 2*. ASCE Geotechnical Special Publication No. 46, American Society of Civil Engineers, New York.

Ahmad, Y.J., S.E. El Serafy and E. Lutz (1989). *Environmental Accounting for Sustainable Development*. The World Bank, Washington, DC.

Alberta Environment and Alberta Labor (1989). Subsurface Remediation Guidelines for Underground Storage Tanks. Alberta MUST (Management of Underground Storage Tanks) Project, A Joint Project of the Departments of Environment and Labor, Edmonton, Alberta, Canada. Draft (August 1989).

Alloway, B.J. and D.C. Ayers (1993). *Chemical Principles of Environmental Pollution*. Blackie Academic & Professional/Chapman & Hall, London, UK.

Ananichev, K. (1976). *Environment: International Aspects*. Progress Publishers, Moscow.

Andelman, J.B. and D.W. Underhill (1988). *Health Effects From Hazardous Waste Sites*, 2nd Printing. Lewis Publishers, Chelsea, Michigan.

Anderson, W.C. (ed.) (1993). *Innovative Site Remediation Technology*. Vols 1–8. American Academy of Environmental Engineers, Annapolis, Maryland.

Arendt, F., G.J. Annokkee, R. Bosman and W.J. van den Brink (eds) (1993). *Contaminated Soil '93, Volumes I & II*. Kluwer Academic Publishers, Dordrecht, The Netherlands.

Asante-Duah, D.K., D.S. Bowles and L.R. Anderson (1991). Framework for the Risk Analysis of Hazardous Waste Facilities. In: *Proceedings of Sixth International Conference on Applications of Statistics and Probability in Civil Engineering*. CERRA/ICASP 6, Mexico.

Ashby, F. (1980). What Price the Furbish Lousewort? *Proc. 4th Conf. on Environ. Engr. Educ.*, Toronto, Canada.

Aswathanarayana, U. (1995). *Geoenvironment: An Introduction*. A.A. Balkema Publishers, Rotterdam, The Netherlands.

Avogadro, A. and R.C. Ragaini (eds) (1994). *Technologies for Environmental Cleanup: Toxic and Hazardous Waste Management*. Kluwer Academic Publishers, Dordrecht, The Netherlands.

Baasel, W.D. (1985). *Economic Methods for Multipollutant Analysis and Evaluation*. Marcel Dekker, Inc., NY.

Barnard, R. and G. Olivetti (1990). Rapid assessment of industrial waste production based on available employment statistics. *Waste Mgmnt Research* **8**(2), 139–144.

Barnhart, M. (1992). Bioremediation: do it yourself. *Soils*, August–Sept., 14–19.

Bartell, S.M., R.H. Gardner and R.V. O'Neill (1992). *Ecological Risk Estimation*. Lewis Publishers, Chelsea, Michigan.

Bates, D.V. (1994). *Environmental Health Risks and Public Policy*. University of Washington Press, Seattle, Washington.

Batstone, R., J.E. Smith, Jr and D. Wilson (eds) (1989). *The Safe Disposal of Hazardous Wastes – The Special Needs and Problems of Developing Countries*, Vols I, II & III. A Joint Study Sponsored by the World Bank, the World Health Organization (WHO), and the United Nations Environment Programme (UNEP). World Bank Technical Paper 0253-7494, No. 93. The World Bank, Washington, DC.

Berthouex, P.M. and L.C. Brown (1994). *Statistics for Environmental Engineers*. Lewis Publishers/CRC Press, Boca Raton, Florida.

Bhatt, H.G., R.M. Sykes and T.L. Sweeney (ed.) (1986). *Management of Toxic and Hazardous Wastes*. Lewis Publishers, Inc., Chelsea, Michigan.

Binder, S., D. Sokal and D. Maughan (1986). Estimating the amount of soil ingested by young children through tracer elements. *Arch. Environ. Health* **41**, 341–345.

Blackman, Jr W.C. (1996). *Basic Hazardous Waste Management*, 2nd edn. Lewis Publishers/CRC Press, Boca Raton, Florida.

Blumenthal, D.S. and A.J. Ruttenber (1995). *Introduction to Environmental Health*, 2nd edn. Springer Publishing Co., NY.

Boulding, J.R. (1994). *Description and Sampling of Contaminated Soils (A Field Guide)*, 2nd edn. Lewis Publishers/CRC Press, Boca Raton, Florida.

Bregman, J.I. and K.M. Mackenthun (1992). *Environmental Impact Statements*. Lewis Publishers, Chelsea, Michigan.

Bretherick, L. (1979). *Handbook of Reactive Chemical Hazards*, 2nd edn. Butterworth Publishers, Wolburn, Massachusetts.

Brown, C. (1978). Statistical aspects of extrapolation of dichotomous dose response data. *J. Nat. Canc. Inst.* **60**, 101–108.

Brown, H.S. (1986). A critical review of current approaches to determining 'How clean is clean' at hazardous waste sites. In: *Hazardous Wastes and Hazardous Materials*, Vol. 3, No. 3, pp. 233–260. Mary Ann Liebert, Inc. Publishers.

Brown, K.W., G.B. Evans, Jr and B.D. Frentrup (eds) (1983). *Hazardous Waste and Treatment*. Butterworth Publishers, Boston, Massachusetts.

Brusick, D.J. (ed.) (1994). *Methods for Genetic Risk Assessment*. Lewis Publishers/CRC Press, Boca Raton, Florida.

Buchel, K.H. (1983). *Chemistry of Pesticides*. John Wiley & Sons, New York.

Byrnes, M.E. (1994). *Field Sampling Methods for Remedial Investigations*. Lewis Publishers/CRC Press, Boca Raton, Florida.

Cairns, J. Jr and T.V. Crawford (ed.) (1991). *Integrated Environmental Management*. Lewis Publishers, Inc., Chelsea, Michigan.

Calabrese, E.J. (1984). *Principles of Animal Extrapolation*. John Wiley & Sons, New York.

Calabrese, E.J. and P.T. Kostecki (1988). *Soils Contaminated by Petroleum: Environment and Public Health Effects*. John Wiley & Sons, New York.

Calabrese, E.J. and P.T. Kostecki (eds) (1989). *Petroleum Contaminated Soils*. Vol. 2. Lewis Publishers, Inc., Chelsea, Michigan.

Calabrese, E.J. and P.T. Kostecki (eds) (1991). *Hydrocarbon Contaminated Soils*, Vol. 1. Lewis Publishers, Inc., Chelsea, Michigan.

Calabrese, E.J. and P.T. Kostecki (1992). *Risk Assessment and Enviornmental Fate Methodologies*. Lewis Publishers/CRC Press, Boca Raton, Florida.

Calabrese, E.J., R. Barnes, E.J. Stanek III, H. Pastides, C.E. Gilbert, P. Veneman, X. Wang, A. Lasztity and P.T. Kostecki (1989). How much soil do young children ingest? An epidemiologic study. *Regulatory Toxicity and Pharmacology* **10**, 123–137.

Canter, L.W., R.C. Knox and D.M. Fairchild (1988). *Ground Water Quality Protection*. Lewis Publishers, Inc., Chelsea, Michigan.

CAPCOA (California Air Pollution Control Officers Association) (1989). *Air Toxics Assessment Manual*. California Air Pollution Control Officers Association, Draft Manual, August 1987 (amended, 1989), California.

Carson, W.H. (ed.) (1990). *The Global Ecology Handbook – What You Can Do About the Environmental Crisis*. The Global Tomorrow Coalition. Beacon Press, Boston, Massachusetts.

Casarett, L.J. and J. Doull (1975). *Toxicology: The Basic Science of Poisons*. Macmillan, New York.

CCME (Canadian Council of Ministers of the Environment) (1991). *Interim Canadian Environmental Quality Criteria for Contaminated Sites*. Report CCME EPC-CS34, The National Contaminated Sites Remediation Program, Winnipeg, Manitoba.

CCME (Canadian Council of Ministers of the Environment) (1993). *Guidance Manual on Sampling, Analysis, and Data Management for Contaminated Sites*. Canadian Council of Ministers of the Environment, Winnipeg, Manitoba.

Chatterji, M. (ed.) (1987). *Hazardous Materials Disposal: Siting and Management*. Gower, Avebury, UK.

Chiu, H.S. and K.L. Tsang (1990). Reduction of treatment cost by using communal treatment facilities. *Waste Mgmnt Research* **8**(2), 165–167.

Chrostowski, P.C., L.J. Pearsall and C. Shaw (1985). Risk assessment as a management tool for inactive hazardous materials disposal sites. *Environ. Mgmnt* **9**(5), 433–442.

Clausing, O., A.B. Brunekreef and J.H. van Wijnen (1987). A method for estimating soil ingestion by children. *Int. Arch. Occup. Environ. Health* **59**, 73–82.

Clayson, D.B., D. Krewski and I. Munro (eds) (1985). *Toxicological Risk Assessment*, Vols 1 and 2. CRC Press, Inc., Boca Raton, Florida.

Clement Associates, Inc. (1988). Multi-Pathway Health Risk Assessment Impact Guidance Document. South Coast Air Quality Management District, California.

Clewell, H.J. and M.E. Andersen (1985). Risk assessment extrapolations and physiological modeling. *Toxicol. Ind. Health* **1**, 111–132.

CMA (Chemical Manufacturers Association) (1984). *Risk Management of Existing Chemicals*. CMA, Washington, DC.

CMA (Chemical Manufacturers Association) (1985). *Risk Analysis in the Chemical Industry*. Government Institutes, Inc., Rockville, Maryland.

Cohen, Y. (1986). Organic pollutant transport. *Environ. Sci. Technol.* **20**(6), 538–545.

Cohrssen, J.J. and V.T. Covello (1989). *Risk Analysis: A Guide to Principles and Methods for Analyzing Health and Environmental Risks*. National Technical Information Service, US Department of Commerce, Springfield, Virginia.

Cole, G.M. (1994). *Assessment and Remediation of Petroleum-contaminated Sites*. Lewis Publishers/CRC Press, Boca Raton, Florida.

Conway, R.A. (ed.) (1982). *Environmental Risk Analysis of Chemicals*. Van Nostrand Reinhold, New York.

Conway, M.F. and S.H. Boutwell (1987). The use of risk assessment to define a corrective action plan for leaking underground storage tanks. In: *Proceedings of the NWWA/API Conference on Petroleum Hydrocarbons & Organic Chemicals in Ground Water – Prevention, Detection & Restoration*. Nov. 1987, pp. 19–40, Water Well Publ. Co., Ohio.

Corn, M. (ed.) (1993). *Handbook of Hazardous Materials*. Academic Press, San Diego, Calif.

Cothern, C.R. (ed.) (1993). *Comparative Environmental Risk Assessment*. Lewis Publishers/CRC Press, Boca Raton, Florida.

Cothern, C.R. and N.P. Ross (eds) (1994). *Environmental Statistics, Assessment, and Forecasting*. Lewis Publishers/CRC Press, Boca Raton, Florida.

Covello, V.T., J. Menkes and J. Mumpower (eds) (1986). *Risk Evaluation and Management. Contemporary Issues in Risk Analysis*, Vol. 1. Plenum Press, New York.

Covello, V.T. and J. Mumpower (1985). Risk analysis and risk management: an historical perspective. *Risk Anal.* **5**, 103–120.

Covello, V.T. et al. (1987). *Uncertainty in Risk Assessment, Risk Management, and Decision Making. Advances in Risk Analysis*, Vol. 4. Plenum Press, New York.

Cowherd, C.M., G.E. Muleski, P.J. Englehart and D.A. Gillette (1984). *Rapid Assessment of Exposure to Particulate Emissions From Surface Contamination Sites*. Midwest Research Institute, Kansas City, Missouri.

Crandall, R.W. and B.L. Lave (eds) (1981). *The Scientific Basis of Risk Assessment*. Brookings Institution, Washington, DC.

Cranor, C.F. (1993). *Regulating Toxic Substances: A Philosophy of Science and the Law*. Oxford University Press, New York.

Crouch, E.A.C. and R. Wilson (1982). *Risk/Benefit Analysis*. Ballinger, Boston, Massachusetts.

Crouch, E.A.C., R. Wilson and L. Zeise (1983). The risks of drinking water. *Water Resources Research* **19**(6), 1359–1375.

Crump, K.S. (1981). An improved procedure for low-dose carcinogenic risk assessment from animal data. *J. Environ. Toxicol.* **5**, 339–346.

Crump, K.S. and R.B. Howe (1984). The multistage model with time-dependent dose pattern: applications of carcinogenic risk assessment. *Risk Analysis* **4**, 163–176.

Csuros, M. (1994). *Environmental Sampling and Analysis for Technicians*. Lewis Publishers/CRC Press, Boca Raton, Florida.

Daniel, D.E. (ed.) (1993). *Geotechnical Practice for Waste Disposal.* Chapman & Hall, London, UK.

Daniels, S.L. (1978). Environmental Evaluation and Regulatory Assessment of Industrial Chemicals. 51st Ann. Conf. Water Poll. Cont. Fed., Anaheim, California.

Davis, A.P. (ed.) (1993). Hazardous and industrial wastes. *Proceedings of the Twenty-Fifth Mid-Atlantic Industrial Waste Conference,* Technomic Publishing Company, Inc., Lancaster, Pennsylvania.

Davis, C.E. (1993). *The Politics of Hazardous Waste.* Prentice-Hall, Englewood Cliffs, New Jersey.

Dawson, G.W. and D. Sanning (1982). Exposure–response analysis for setting site restoration criteria. *Proc. National Conf. on Mgmnt of Uncontrolled Hazardous Waste Sites,* Washington, DC.

Deuel, L.E. Jr and G.H. Holliday (1994). *Soil Remediation for the Petroleum Extraction Industry.* Pennwell Books/Pennwell Publishing Company, Tulsa, Oklahoma.

Devinny, J.S., L.G. Everett, J.C.S. Lu and R.L. Stollar (1990). *Subsurface Migration of Hazardous Wastes.* Van Nostrand Reinhold, New York.

Diesler, P.F. (ed.) (1984). *Reducing the Carcinogenic Risks in Industry.* Marcel Dekker, New York.

Doherty, N., P. Kleindorfer and H. Kunreuther (1990). An insurance perspective on an integrated waste management strategy. In: H. Kunreuther and M.V. Rajeer Gowda (eds), *Integrating Insurance and Risk Management for Hazardous Wastes.* Kluwer Academic Publishers, Boston, Massachusetts, pp. 271–302.

de Serres, F.J. and A.D. Bloom (eds) (1996). *Ecotoxicity and Human Health (A Biological Approach to Environmental Remediation).* CRC Press/Lewis Publishers, Boca Raton, Florida, USA.

Dunster, H.J. and W. Vinck (1979). The Assessment of Risk, its Value and Limitations. European Nuclear Conference, Hamburg, Germany.

Ellis, B. and J.F. Rees (1995). Contaminated land remediation in the UK with reference to risk assessment: two case studies. *J. Inst. Water Environ. Manag.* 9(1), February, 27–36.

Erickson, A.J. (1977). *Aids for Estimating Soil Erodibility – 'K' Value Class and Tolerance.* US Dept. of Agriculture, Soil Conservation Service, Salt Lake City, Utah.

Erickson, M.D. (1993). *Remediation of PCB Spills.* Lewis Publishers/CRC Press, Boca Raton, Florida.

Erickson, R.L. (1992). *Environmental Remediation Contracting.* John Wiley & Sons, New York.

Eschenroeder, A., R.J. Jaeger, J.J. Ospital and C. Doyle (1986). Health risk assessment of human exposure to soil amended with sewage sludge contaminated with polychlorinated dibenzodioxins and dibenzofurans. *Vet. Hum. Toxicol.* 28, 356–442.

Ess, T.H. (1981a) Risk acceptability. In: *Proceedings of National Conference on Risk & Decision Analysis for Hazardous Waste Disposal,* 24–27 Aug., Baltimore, Maryland, pp. 164–174.

Ess, T.H. (1981b). Risk estimation. In: *Proceedings of National Conference on Risk & Decision Analysis for Hazardous Waste Disposal,* 24–27 Aug., Baltimore, Maryland, pp. 155–163.

Evans, L.J. (1989). Chemistry of metal retention by soils. *Environ. Sci. Technol.* 23(9), 1047–1056.

Fetter, C.W. (1993). *Contaminant Hydrogeology.* Macmillan, New York.

Fitchko, J. (1989). *Criteria for Contaminated Soil/Sediment Cleanup.* Pudvan Publishing Co., Northbrook, Illinois.

Forester, W.S. and J.H. Skinner (eds) (1987). *International Perspectives on Hazardous Waste Management – A Report from the International Solid Wastes and Public Cleansing Association (ISWA) Working Group on Hazardous Wastes.* Academic Press, London, UK.

FPC (Florida Petroleum Council) (1986). *Benzene in Florida Groundwater: An Assessment of the Significance to Human Health.* Florida Petroleum Council, Florida.

Francis, B.M. (1994). *Toxic Substances in the Environment.* John Wiley & Sons, New York.

Fredrickson, J.K., H. Bolton, Jr and F.J. Brockman (1993). In situ and on-site bioreclamation. *Environ. Sci. Technol. (ES&T)* 27(9), 1711–1716.

Freeman, H.M. and E.F. Harris (eds) (1995). *Hazardous Waste Remediation: Innovative Treatment Technologies.* Technomic Publishing Co., Inc., Lancaster, Pennsylvania.

Freese, E. (1973). Thresholds in toxic, tetratogenic, mutagenic, and carcinogenic effects. *Environ. Health Pers.* 6, 171–178.

Garrett, P. (1988). *How to Sample Groundwater and Soils.* National Water Well Association (NWWA), Dublin, Ohio.

Gilpin, A. (1995). *Environmental Impact Assessment (EIA): Cutting Edge for the Twenty-first Century*. Cambridge University Press, Cambridge, UK.

Glasson, J., R. Therivel and A. Chadwick (1994). *Introduction to Environmental Impact Assessment*. University College London, UCL Press, London, UK.

Glickman, T.S. and M. Gough (eds) (1990). *Readings in Risk*. Resources for the Future, Washington, DC.

Goldman, S.J., K. Jackson and T.A. Bursztynsky (1986). *Erosion and Sediment Control Handbook*. McGraw-Hill, New York.

Gordon, S.I. (1985). *Computer Models in Environmental Planning*. Van Nostrand Reinhold, New York.

Gorelick, S.M., R.A. Freeze, D. Donohue and J.F. Keely (1993). *Groundwater Contamination (Optimal Capture and Containment)*. Lewis Publishers/CRC Press, Boca Raton, Florida.

Gots, R.E. (1993). *Toxic Risks*. Lewis Publishers/CRC Press, Boca Raton, Florida.

Grasso, D. (1993). *Hazardous Waste Site Remediation (Source Control)*. Lewis Publishers/CRC Press, Boca Raton, Florida.

Gregory, R. and H. Kunreuther (1990). Successful siting incentives. *Civil Engineering* **60**(4), 73–75.

Grisham, J.W. (ed.) (1986). *Health Aspects of the Disposal of Waste Chemicals*. Pergamon Press, Oxford, England.

Gunn J.M. (ed.) (1995). *Restoration and Recovery of an Industrial Region: Progress in Restoring the Smelher-Damaged Landscape Near Sudbury, Canada*. Springer Series on Environmental Management, Springer-Verlag, New York.

Guswa J.H., W.J. Lyman, A.S. Donigian, Jr, T.Y.R. Lo and E.W. Stanahan (1984). *Groundwater Contamination and Emergency Response Guide*. Noyes Publications, Park Ridge, New Jersey.

Haas, C.N. and R.J. Vamos (1995). *Hazardous and Industrial Waste Treatment*. Prentice-Hall, Englewood Cliffs, New Jersey.

Hadley, P.W. and R.M. Sedman (1990). A health-based approach for sampling shallow soils at hazardous waste sites using the $AAL_{soil\ contact}$ criterion. *Environmental Health Perspectives* **18**, 203–207.

Haimes, Y.Y., L. Duan and V. Tulsiani (1990). Multiobjective decision-tree analysis. *Risk Analysis* **10**(1), 111–129.

Haith, D.A. (1980). A mathematical model for estimating pesticide losses in runoff. *J. Environ. Qual.* **93**(3), 428–433.

Hallenbeck, W.H. and K.M. Cunningham-Burns (1985). *Pesticides and Human Health*. Springer-Verlag, New York.

Hansen, P.E. and S.E. Jorgensen (eds) (1991). Introduction to environmental management. *Developments in Environmental Modelling*, **18**, Elsevier, Amsterdam, The Netherlands.

Hasan, S.E. (1996). *Geology and Hazardous Waste Management*. Prentice-Hall, Upper Saddle River, New Jersey.

Haun, J.W. (1991). *Guide to the Management of Hazardous Waste*. Fulcrum Publishing, Golden, Colorado.

Hawken, P. (1993). *The Ecology of Commerce: A Declaration of Sustainability*. Harper Business, New York.

Hawley, G.G. (1981). *The Condensed Chemical Dictionary*, 10th edn. Van Nostrand Reinhold, New York.

Hawley, J.K. (1985). Assessment of health risks from exposure to contaminated soil. *Risk Analysis* **5**(4), 289–302.

Hayes, A.W. (ed.) (1982). *Principles and Methods of Toxicology*. Raven Press, New York.

Henderson, M. (1987). *Living with Risk: The Choices, The Decisions*. British Medical Association, Somerset, NJ/John Wiley & Sons, New York.

Hickey, R.F. and G. Smith (eds) (1996). *Biotechnology in Industrial Waste Treatment and Bioremediation*. Lewis Publishers/CRC Press, Boca Raton, Florida.

Higgins, T.E. (ed.) (1995). *Pollution Prevention Handbook*. Lewis Publishers/CRC Press, Boca Raton, Florida.

Hinchee, H.J., H.J. Reinsinger, D. Burris, B. Marks and J. Stepek (1986). Underground fuel contamination, investigation, and remediation: a risk assessment approach to how clean is clean. In: *Proceedings of the NWWA/API Conference on Petroleum Hydrocarbons & Organic Chemicals in Ground Water – Prevention, Detection & Restoration*, Nov. 1986, Water Well Publ. Co., Ohio, pp. 539–564.

Hoddinott, K.B. (ed.) (1992). *Superfund Risk Assessment in Soil Contamination Studies*. American Society for Testing and Materials, ASTM Publication STP 1158, Philadelphia, Pennsylvania.

Hoel, D.G., D.W. Gaylor, R.L. Kirschstein, U. Saffiotti and M.A. Schneiderman (1975). Estimation of risks of irreversible, delayed toxicity. *J. Toxicol. Environ. Health* **1**, 133–151.

Honeycutt, R.C. and D.J. Schabacker (eds) (1994). *Mechanisms of Pesticide Movement into Groundwater*. Lewis Publishers/CRC Press, Boca Raton, Florida.

HSE (Health and Safety Executive) (1989a). *Risk Criteria for Land-Use Planning in the Vicinity of Major Industrial Hazards*. HMSO, London, UK.

HSE (Health and Safety Executive) (1989b). *Quantified Risk Assessment – Its Input to Decision Making*. HMSO, London, UK.

Huang, C.P. (ed.) (1994). *Hazardous and Industrial Wastes. Proceedings of the Twenty-Sixth Mid-Atlantic Industrial Waste Conference*, Technomic Publishing Company, Inc., Lancaster, Pennsylvania.

Hunt, J.R., J.T. Geller, N. Sitar and K.S. Udell (1988). *Subsurface Transport Processes for Gasoline Components*. In: Specialty Conf. Proc., Joint CSCE-ASCE National Conf. on Environ. Engr., Vancouver, DC, Canada, 13–15 July 1988, pp. 536–543.

Hwang, S.T. and J.W. Falco (1986). Estimation of multimedia exposures related to hazardous waste facilities. In: Y. Cohen (ed.). *Pollutants in a Multimedia Environment*. Plenum Press, New York.

IARC (International Agency for Research on Cancer) (1984). *IARC Monographs on the Evaluation of the Carcinogenic Risk of Chemicals to Humans*, Vol. 33. World Health Organization, Lyons, France.

ICRCL (Interdepartmental Committee on the Redevelopment of Contaminated Land) (1987). *Guidance on the Assessment and Redevelopment of Contaminated Land*. ICRCL 59/83, 2nd edn. Department of the Environment, Central Directorate on Environmental Protection, London, UK.

IJC (International Joint Commission) (1986). *Literature Review of the Effects of Persistent Toxic Substances on Great Lakes Biota*. Report of the Health of Aquatic Communities Task Force, International Joint Commission.

IRPTC (1978). *Attributes for the Chemical Data Register of the International Register of Potentially Toxic Chemicals*. Register Attribute Series, No. 1. International Register of Potentially Toxic Chemicals, UNEP, Geneva, Switzerland.

IRPTC (1985). *International Register of Potentially Toxic Chemicals*, Part A. International Register of Potentially Toxic Chemicals, UNEP, Geneva, Switzerland.

IRPTC (1985). *Industrial Hazardous Waste Management*. Industry and Environment Office and the International Register of Potentially Toxic Chemicals. United Nations Environment Programme, Geneva, Switzerland.

Jain, R.K., L.V. Urban, G.S. Stacey and H.E. Balbach (1993). *Environmental Assessment*. McGraw-Hill, New York.

J.C. Consultancy Ltd, London (1986). *Risk Assessment for Hazardous Installations*. Pergamon Press, Oxford, England.

Jennings, A.A. and P. Suresh (1986). Risk penalty functions for hazardous waste management. *ASCE, J. Environ. Engnr.* **112**(1), February, 105–122.

Johnson, B.B. and V.T. Covello (1987). *Social and Cultural Construction of Risk: Essays on Risk Selection and Perception*. Kluwer Academic Publishers, Norwell, Massachusetts.

Jorgensen, E.P. (ed.) (1989). *The Poisoned Well – New Strategies for Groundwater Protection*. Sierra Club Legal Defense Fund. Island Press, Washington, DC.

Kastenberg, W.E., T.E. McKone and D. Okrent (1976). *On Risk Assessment in the Absence of Complete Data*. UCLA Rep. No. UCLA-ENG-677, School of Engineering & Applied Science, California.

Kastenberg, W.E. and H.C. Yeh (1993). Assessing public exposure to pesticides – contaminated ground water. *J. Ground Water* **31**(5), 746–752.

Kates, R.W. (1978). *Risk Assessment of Environmental Hazard*. SCOPE Report 8. John Wiley & Sons, Chichester, UK.

Keeney, R.D. and H. Raiffa (1976). *Decisions with Multiple Objectives: Preferences and Value Tradeoffs*. John Wiley & Sons, New York.

Keeney, R.L. (1984). Ethics, decision analysis, and public risk. *Risk Anal.* **4**, 117–129.

Keeney, R.L. (1990). Mortality risks induced by economic expenditures. *Risk Anal.* **10**(1), 147–159.

Keith, L.H. (ed.) (1992). *Compilation of E.P.A.'s Sampling and Analysis Methods*. Lewis Publishers/CRC Press, Boca Raton, Florida.

Kempa, E.S. (ed.) (1991). *Environmental Impact of Hazardous Wastes*. PZITS Publishing Dept., Warsaw, Poland.

Kimmel, C.A. and D.W. Gaylor (1988). Issues in qualitative and quantitative risk analysis for developmental toxicology. *Risk Anal.* **8**, 15–20.

Kleindorfer, P.R. and H.C. Kunreuther (eds) (1987). *Insuring and Managing Hazardous Risks: From Seveso to Bhopal and Beyond*. Springer-Verlag, Berlin, Germany.

Knowles, P-C. (ed.) (1992). *Fundamentals of Environmental Science and Technology*. Government Institutes, Inc., Rockville, Maryland.

Kostecki, P.T. and E.J. Calabrese (eds) (1989). *Petroleum Contaminated Soils*, Vols 1, 2, and 3. Lewis Publishers, Inc., Chelsea, Michigan.

Kostecki, P.T. and E.J. Calabrese (eds) (1991). *Hydrocarbon Contaminated Soils and Ground-water*. Vol. 1. Lewis Publishers, Inc., Chelsea, Michigan.

Kostecki, P.T. and E.J. Calabrese (eds) (1992). *Contaminated Soils (Diesel Fuel Contamination)*. Lewis Publishers/CRC Press, Boca Raton, Florida.

Kreith, F. (ed.) (1994). *Handbook of Solid Waste Management*. McGraw-Hill, New York.

Krewski, D., C. Brown and D. Murdoch (1984). Determining safe levels of exposure: safety factors or mathematical models. *Fundam. Appl. Toxicol.* **4**, S383–S394.

Kunreuther, H. and M.V. Rajeev Gowda (eds) (1990). *Integrating Insurance and Risk Management for Hazardous Wastes*. Kluwer Academic Publishers, Boston, Massachusetts.

LaGoy, P.K. (1987). Estimated soil ingestion rates for use in risk assessment. *Risk Anal.* **7**(3), 355–359.

Larsen, R.J. and M.L. Marx (1985). *An Introduction to Probability and its Applications*. Prentice-Hall, Englewood Cliffs, New Jersey.

Lave, L.B. (ed.) (1982). *Quantitative Risk Assessment in Regulation*. The Brooking Institute, Washington, DC.

Lave, L.B. and A.C. Upton (eds) (1987). *Toxic Chemicals, Health, and the Environment*. The Johns Hopkins University Press, Baltimore, Maryland.

Lee, J.A. (1985). *The Environment, Public Health, and Human Ecology – Considerations for Economic Development*. A World Bank Publication. The Johns Hopkins University Press, Baltimore, Maryland.

Lees, F.B. (1980). *Loss Prevention in the Process Industries*, Vol. 1. Butterworths, Boston, Massachusetts.

Lepow, M.L., M. Bruckman, L. Robino, S. Markowitz, M. Gillette and J. Kapish (1974). Role of airborne lead in increased body burden of lead in Hartford children. *Environ. Health Persp.* **6**, 99–101.

Lepow, M.L., M. Bruckman, M. Gillette, S. Markowitz, R. Robino and J. Kapish (1975). Investigations into sources of lead in the environment of urban children. *Environ. Res.* **10**, 415–426.

Lifson, M.W. (1972). *Decision and Risk Analysis for Practicing Engineers*. Barnes and Noble, Cahners Bks., Boston, Massachusetts.

Lind, N.C., J.S. Nathwani and E. Siddall (1991). *Managing Risks in the Public Interest*. Institute for Risk Research, Univ. of Waterloo, Waterloo, Ontario.

Lindsay, W.L. (1979). *Chemical Equilibria in Soils*. Wiley-Interscience, New York.

Linthurst, R.A., P. Bourdeau and R.G. Tardiff (eds) (1995). Methods to assess the effects of chemicals on ecosystems. *SCOPE 53/IPCS Joint Activity 23/SGOMSEC 10*, J. Wiley & Sons, Chichester, UK.

Lippit, J., J. Walsh, M. Scott and A. DiPuccio (1986). *Cost of Remedial Actions at Uncontrolled Hazardous Waste Sites: Worker Health and Safety Considerations*. Project Summary Report No. EPA/600/S2-86/037. US EPA, Washington, DC.

Lippmann, M. (ed.) (1992). *Environmental Toxicants: Human Exposures and Their Health Effects*. Van Nostrand Reinhold, New York.

Liptak, S.C., J.W. Atwater and D.S. Mavinic (eds) (1988). *Proceedings of the 1988 Joint CSCE– ASCE National Conference on Environmental Engineering*. 13–15 July, Pan Pacific Hotel, Vancouver, BC, Canada.

Loehr, R.C. (ed.) (1976). *Land as a Waste Management Alternative*. Ann Arbor Science Publ., Inc., Ann Arbor, Michigan.

Long, W.L. (1990). Economic aspects of transport and disposal of hazardous wastes. *Marine Policy International Journal* **14**(3), May, 198–204.

Long, F.A. and G.E. Schweitzer (eds) (1982). *Risk Assessment at Hazardous Waste Sites*. American Chemical Society, Washington, DC.

Lowrance, W.W. (1976). *Of Acceptable Risk: Science and the Determination of Safety*. William Kaufman, Inc., Los Altos, California.

Lu, F.C. (1985). *Basic Toxicology*. Hemisphere, Washington, DC.

Lu, F.C. (1985). Safety assessments of chemicals with threshold effects. *Regul. Toxicol. Pharmacol.* **8**, 121–132.

Lu, F.C. (1988). Acceptable daily intake: inception, evolution, and application. *Regul. Toxicol. Pharmacol.* **8**, 45–60.

LUFT (1989). *Leaking Underground Fuel Tank Field Manual: Guidelines for Site Assessment, Cleanup, and Underground Storage Tank Closure*. State of California, Leaking Underground Fuel Tank Task Force, October 1989.

MacCarthy, L.S. and D. Mackay (1993). Enhancing ecotoxicological modeling and assessment. *Environ. Sci. Technol. (ES&T)* **27**(9), 1719–1728.

MacDonald, J.A. and B.E. Rittmann (1993). Performance standards for in situ bioremediation. *Environ. Sci. Technol (ES&T)* **27**(10), 1974–1979.

Mackay, D. and P.J. Leinonen (1975). Rate of evaporation of low-solubility contaminants from water bodies. *Environ. Sci. Technol.* **9**, 1178–1180.

Mackay, D. and A.T.K. Yeun (1983). Mass transfer coefficient correlations for volatilization of organic solutes from water. *Environ. Sci. Technol.* **17**, 211–217.

Macari, E.J., J.D. Frost and L.F. Pumarada (eds) (1995). Geo-environmental issues facing the Americas. *ASCE Geotechnical Special Publication No. 47*, American Society of Civil Engineers, New York.

Mahmood, R.J. and R.C. Sims (1986). Mobility of organics in land treatment systems. *J. Environ. Engnr.* **112**(2), 236–245.

Malina, Jr, J.F. (ed.) (1989). *Environmental Engineering: Proceedings of the 1989 Specialty Conference*, Austin, Texas, 10–12 July 1989. ASCE, New York.

Manahan, S. (1992). *Toxicological Chemistry*. Lewis Publishers/CRC Press, Boca Raton, Florida.

Manahan, S.E. (1993). *Fundamentals of Environmental Chemistry*. Lewis Publishers/CRC Press, Boca Raton, Florida.

Mansour, M. (ed.) (1993). *Fate and Prediction of Environmental Chemicals in Soils, Plants, and Aquatic Systems*. Lewis Publishers/CRC Press, Boca Raton, Florida.

Martin, E.J. and J.H. Johnson, Jr (eds) (1987). *Hazardous Waste Management Engineering*. Van Nostrand Reinhold, New York.

Martin, W.F., J.M. Lippitt and T.G. Prothero (1992). *Hazardous Waste Handbook for Health and Safety*, 2nd edn. Butterworth-Heinemann, London, UK.

Massmann, J. and R.A. Freeze (1987). Groundwater contamination from waste management sites: the interaction between risk-based engineering design and regulatory policy, 1, Methodology; & 2, Results. *Water Resour. Res.* **23**(2), 351–380.

Mathews, J.T. (1991). *Preserving the Global Environment: The Challenge of Shared Leadership*. W.W. Norton & Co, New York.

Maurits la Riviere, J.W. (1989). Threats to the world's water. *Scientific American*. Managing Planet Earth. Sept. 1989, Special Issue.

McColl, R.S. (ed.) (1987). *Environmental Health Risks: Assessment and Management*. Institute for Risk Research, University of Waterloo Press, Waterloo, Ontario, Canada.

McKone, T.E. and J.I. Daniels (1991). Estimating human exposure through multiple pathways from air, water, and soil. *Regul. Toxicol. Pharmacol.* **13**, 36–61.

McKone, T.E., W.E. Kastenberg and D. Okrent (1983). The use of landscape chemical cycles for indexing the health risks of toxic elements and radionuclides. *Risk Anal.* **3**(3), 189–205.

McTernan, W.F. and E. Kaplan (eds) (1990). *Risk Assessment for Groundwater Pollution Control*. American Society of Civil Engineers, New York.

Merck (1989). *The Merck Index: An Encyclopedia of Chemicals, Drugs and Biologicals*, 11th (Centennial) edn. Merck & Co., Inc., Rockway, New Jersey.

Meyer, C.R. (1983). Liver dysfunction in residents exposed to leachate from a toxic waste dump. *Environ. Health Pers.* **48**, 9–13.

Miller, D.W. (ed.) (1980). *Waste Disposal Effects on Ground Water*. Premier Press, Berkeley, California.

Mills, W.B., Dean, J.D., Procella, D.B. et al. (1982). *Water Quality Assessment: a Screening Procedure for Toxic and Conventional Pollutants*: parts 1, 2, and 3. Athens, GA: US Environmental Protection Agency. Environmental Research Laboratory. Office of Research and Development. EPA-600/6-82/004 a.b.c.

Mitchell, J.K. and G.D. Bubenzer (1980). Soil loss estimation. In: M.J. Kirby and R.P.C. Morgan (eds), *Soil Erosion*. John Wiley & Sons, New York.

Mitchell, J.K. (1993). *Fundamentals of Soil Behavior, 2nd edition*. J. Wiley & Sons, New York.

Mitsch, W.J. and S.E. Jorgesen (eds) (1989). *Ecological Engineering, An Introduction into Ecotechnology*. John Wiley & Sons, New York.

Mockus, J. (1972). Estimation of direct runoff from storm rainfall. In: *National Engineering Handbook. Section 4: Hydrology*. US Department of Agriculture Soil Conservation Service, Washington, DC.

Moeller, D.W. (1992). *Environmental Health*. Havard University Press, Cambridge, Mass.

Monahan, D.J. (1990). Estimation of hazardous wastes from employment statistics – Victoria, Australia. *Waste Mgmnt Research* **8**(2), 145–149.

Moore, A.O. (1987). *Making Polluters Pay – A Citizen's Guide to Legal Action and Organizing*. Environmental Action Foundation, Washington, DC.

Morris, P. and R. Therivel (eds) (1995). *Methods of Environmental Impact Assessment*. UCL Press, University College London, UK.

Mulkey, L.A. (1984). Multimedia fate and transport models: an overview. *J. Toxicol. – Clin. Toxicol.* **21**(1–2), 65–95.

Munro, I.C. and D.R. Krewski (1981). Risk assessment and regulatory decision making. *Food. Cosmet. Toxicol.* **19**, 549–560.

NAE (National Academy of Engineering). (1993). *Keeping Pace with Science and Engineering: Case Studies in Environmental Regulation*. National Academy Press, Washington, DC.

Nathwani, J., N.C. Lind and E. Siddall (1990). Risk–benefit balancing in risk management: measures of benefits and detriments. Presented at the Annual Meeting of the Society for Risk Analysis, 29 Oct.–1 Nov. 1989, San Francisco, California. Institute for Risk Research Paper No. 18, Waterloo, Ontario, Canada.

Neely, W.B. (1980). *Chemicals in the Environment (Distribution, Transport, Fate, Analysis)*. Marcel Dekker, Inc., New York.

Neely, W.B. (1994). *Introduction to Chemical Exposure and Risk Assessment*. Lewis Publishers/ CRC Press, Boca Raton, Florida.

Nielsen, D.M. (ed.) (1991). *Practical Handbook of Groundwater Monitoring*. Lewis Publishers, Chelsea, Michigan.

NIOSH (National Institute for Occupational Safety and Health) (1982). *Registry of Toxic Effects of Chemical Substances*. R.L. Tatken and R.J. Lewis (eds). US Dept. Health and Human Services (DDSH). NIOSH. Cincinnati, OH DHHS (NIOSH) Publ. No. 83–107.

Norris, R.D. et al. (1994). *Handbook of Bioremediation*. Lewis Publishers/CRC Press, Boca Raton, Florida.

NRC (National Research Council) (1972). *Specifications and Criteria for Biochemical Compounds*, 3rd edn. National Academy of Sciences, Washington, DC.

NRC (National Research Council) (1977). *Drinking Water and Health*, Vol. 1. National Academy Press, Washington, DC.

NRC (National Research Council) (1977). *Environmental Monitoring, Analytical Studies for the US Environmental Protection Agency*, Vol. IV. National Academy Press, Washington, DC.

NRC (National Research Council) (1977). *Drinking Water and Health*. Safe Drinking Water Committee, Advisory Center on Toxicology, National Academy of Sciences, Washington, DC.

NRC (National Research Council) (1980). *Drinking Water and Health*, Vol. 2. National Academy Press, Washington, DC.

NRC (National Research Council) (1980). *Drinking Water and Health*, Vol. 3. National Academy Press, Washington, DC.

NRC (National Research Council) (1981). *Prudent Practices for Handling Hazardous Chemicals in Laboratories*. National Academy Press, Washington, DC.

NRC (National Research Council) (1982). *Drinking Water and Health*, Vol. 4. National Academy Press, Washington, DC.

NRC (National Research Council) (1982). *Risk and Decision-Making: Perspective and Research*. Committee on Risk and Decision-Making. National Academy Press, Washington, DC.

NRC (National Research Council) (1983). *Drinking Water and Health*, Vol. 5. National Academy Press, Washington, DC.

NRC (National Research Council) (1988). *Hazardous Waste Site Management: Water Quality Issues*. National Academy Press, Washington, DC.

NRC (National Research Council) (1989). *Ground Water Models: Scientific and Regulatory Applications*. National Academy Press, Washington, DC.

NRC (National Research Council) (1991). *Environmental Epidemiology (Public Health and Hazardous Wastes)*. National Academy Press, Washington, DC.

NRC (National Research Council) (1993). *Ground Water Vulnerability Assessment: Predicting Relative Contamination Potential Under Conditions of Uncertainty*. National Academy Press, Washington, DC.

NRC (National Research Council) (1993). *In Situ Bioremediation: When Does it Work?* National Academy Press, Washington, DC.

NRC (National Research Council) (1993). *Issues in Risk Assessment*. National Academy Press, Washington, DC.

NRC (National Research Council) (1994). *Alternatives for Ground Water Cleanup*. Committee on Ground Water Cleanup Alternatives. National Academy Press, Washington, DC.

NRC (National Research Council) (1994). *Ranking Hazardous Waste Sites for Remedial Action*. National Academy Press, Washington, DC.

NRC (National Research Council) (1995). *Improving the Environment: An Evaluation of DOE's Environmental Management Program*. National Academy Press, Washington, DC.

OECD (1986). *Existing Chemicals – Systematic Investigation, Priority Setting and Chemical Review*. Organization for Economic Cooperation and Development, Paris, France.

OECD (1986). *Report of the OECD Workshop on Practical Approaches to the Assessment of Environmental Exposure*. 14–18 April, Vienna, Austria.

OECD (1994). *Environmental Indicators: OECD Core Set*. Organization for Economic Cooperation and Development, Paris, France.

O'Hare, M., L. Bacow and D. Sanderson (1983). *Facility Sitting and Public Opposition*. Van Nostrand Reinhold, New York.

Onishi, Y., A.R. Olsen, M.A. Parkhurst and G. Whelan (1985). Computer-based environmental exposure and risk assessment methodology for hazardous materials. *J. Hazard. Mater.* **10**, 389–417.

OSHA (Occupational Safety and Health Administration) (1980). Identification, classification, and regulation of potential occupational carcinogens. *Federal Register* **45**, 5002–5296.

O'Shay, T.A. and K.B. Hoddinott (eds) (1994). *Analysis of Soils Contaminated with Petroleum Constituents*. ASTM Publication, STP 1221, ASTM, Philadelphia, Pennsylvania.

OSTP (Office of Science and Technology Policy) (1985). Chemical carcinogens: a review of the science and its associated principles. *Federal Register* **50**, 10372–10442.

OTA (Office of Technology Assessment) (1981). *Assessment of Technologies for Determining Cancer Risks from the Environment*. Office of Technology Assessment, Washington, DC.

OTA (Office of Technology Assessment) (1983). *Technologies and Management Strategies for Hazardous Waste Control.* Congress of the US Office of Technology Assessment, Washington, DC.

Ott, W.R. (1995). *Environmental Statistics and Data Analysis.* Lewis Publishers/CRC Press, Boca Raton, Florida.

Overcash, M.R. and D. Pal (1979). *Design and Land Treatment Systems for Industrial Waste – Theory and Practice.* Ann Arbor Science Publ., Inc., Ann Arbor, Michigan.

Oweis, I.S. and R.P. Khera (1990). *Geotechnology of Waste Management.* Butterworths, London, UK.

Paasivira, J. (1991). *Chemical Ecotoxicology.* Lewis Publishers, Chelsea, Michigan.

Park, C.N. and R.D. Snee (1983). Quantitative risk assessment: state of the art for carcinogenesis. *Am. Stat.* **37**(4), 427–441.

Patnaik, P. (1992). *A Comprehensive Guide to the Hazardous Properties of Chemical Substances.* Van Nostrand Reinhold, New York.

Peck, D.L. (ed.) (1989). *Psychosocial Effects of Hazardous Toxic Waste Disposal in Communities.* Charles C. Thomas Publishers, Springfield, Illinois.

Pedersen, J. (1989). *Public Perception of Risk Associated with the Siting of Hazardous Waste Treatment Facilities.* European Foundation for the Improvement of Living and Working Conditions, Dublin, Eire.

Perket, C.L. (ed.) (1986). *Quality Control in Remedial Site Investigation: Hazardous and Industrial Solid Waste Testing,* Vol. 5. American Society for Testing and Materials (ASTM) STP 925, ASTM, Philadelphia, Pennsylvania.

Petak, W.J. and A.A. Atkisson (1982). *Natural Hazard Risk Assessment and Public Policy: Anticipating the Unexpected.* Springer-Verlag, New York.

Philp, R.B. (1995). *Environmental Hazards and Human Health.* Lewis Publishers/CRC Press, Boca Raton, Florida.

Pickering, Q.H. and C. Henderson (1966). The acute toxicity of some heavy metals to different species of warm water fishes. *Air Water Pollut. Int. J.* **10**(6/7), 453–463.

Piddington, K.W. (1989). Sovereignty and the environment. *Environment* **31**(7), September, 18–20, 35–39.

Pierzynski, G.M., J.T. Sims and G.F. Vance (1994). *Soils and Environmental Quality.* Lewis Publishers/CRC Press, Boca Raton, Florida.

Postel, S. (1988). *Controlling Toxic Chemicals. State of the World 1988.* Worldwatch Institute, New York.

Prager, J.C. (1995). *Environmental Contaminant Reference Databook.* Van Nostrand Reinhold, New York.

Prins, G. and R. Stamp (1991). *Top Guns and Toxic Whales: The Environment and Global Security.* Earthscan Publications, London, UK.

Purdue University (1990). *Proceedings of the 44th Industrial Waste Conference* (9–11 May 1989). Purdue Univ., West Lafayette, Indiana. Lewis Publishers, Chelsea, Michigan.

Purdue University (1991). *Proceedings of the 45th Industrial Waste Conference* (8–10 May 1990). Purdue Univ., West Lafayette, Indiana. Lewis Publishers, Chelsea, Michigan.

Purdue University (1992). *Proceedings of the 46th Industrial Waste Conference* (May 1991). Purdue Univ., West Lafayette, Indiana. Lewis Publishers, Chelsea, Michigan.

Purdue University (1993). *Proceedings of the 47th Industrial Waste Conference* (May 1992). Purdue Univ., West Lafayette, Indiana. Lewis Publishers, Chelsea, Michigan.

Rail, C.C. (1989). *Groundwater Contamination: Sources, Control and Preventive Measures.* Technomic Publishing Co., Inc., Lancaster, Pennsylvania.

Ramamoorthy, S. and E. Baddaloo (1991). *Evaluation of Environmental Data for Regulatory and Impact Assessment. Studies in Environmental Science 41.* Elsevier Science, Amsterdam, The Netherlands.

Ray, D.L. (1990). *Trashing the Planet.* Regnery Gateway, Washington, DC.

Reed, S.C., R.W. Crites and E.J. Middlebrooks (1995). *Natural Systems for Waste Management and Treatment, 2nd edn.* McGraw-Hill, New York.

Reeve, R.N. (1994). *Environmental Analysis.* John Wiley & Sons, New York.

Rhyner, C.R., L.J. Schwartz, R.B. Wenger and M.G. Kohrell (1995). *Waste Management and Resource Recovery.* Lewis Publishers/CRC Press, Boca Raton, Florida.

Ricci, P.F. (ed.) (1985). *Principles of Health Risk Assessment*. Prentice-Hall, Englewood Cliffs, New Jersey.

Ricci, P.F. and M.D. Rowe (eds) (1985). *Health and Environmental Risk Assessment*. Pergamon Press, New York.

Richardson, M.L. (ed.) (1992). *Risk Management of Chemicals*. Royal Society of Chemicals, Cambridge, UK.

Rodricks, J.V. (1984). Risk assessment at hazardous waste disposal sites. *Hazardous Waste and Hazardous Materials* 1(3), 333–362.

Rodricks, J.V. and R.G. Tardiff (eds) (1984). *Assessment and Management of Chemical Risks*. ACS Symposium Series 239, American Chemical Society, Washington, DC.

Rodricks, J. and Taylor, M.R. (1983). Application of risk assessment to food safety decision making. *Regulatory Toxicol. Pharmacol.* 3, 275–307.

Rowe, W.D. (1983). *Evaluation Methods for Environmental Standards*. CRC Press, Inc., Boca Raton, Florida.

Rowland, A.J. and P. Cooper (1983). *Environment and Health*. Edward Arnold, England.

Royal Society of London (1983). *Risk Assessment: A Study Group Report*. The Royal Society, London.

Ruckelshaus, W.D. (1985). Risk, science, and democracy. *Issues Sci. Technol.* spring, 19–38.

Russell, M. and M. Gruber (1987). Risk assessment in environmental policy-making. *Science* **236**, 286–290.

Santos, S.L. and J. Sullivan (1988). The use of risk assessment for establishing corrective action levels at RCRA sites. In *Hazardous Wastes and Hazardous Materials*. Mary Ann Liebert, Inc., Publishers.

Sara, M.N. (1993). *Standard Handbook of Site Assessment for Solid and Hazardous Waste Facilities*. Lewis Publishers/CRC Press, Boca Raton, Florida.

Sax, N.I. (1979). *Dangerous Properties of Industrial Materials*, 5th edn. Van Nostrand Reinhold, New York.

Saxena, J. and F. Fisher (eds) (1981). *Hazard Assessment of Chemicals*. Academic Press, New York.

Schleicher, K. (ed.) (1992). *Pollution Knows No Frontiers: A Reader*. Paragon House Publishers, New York.

Schramm, G. and J.J. Warford (eds). *Environmental Management and Economic Development*. A World Bank Publication. The Johns Hopkins University Press, Baltimore, Maryland.

Schroeder, R.L. (1985). *Habitat Suitability Index Models: Northern Bobwhite*. US Department of Interior, Fish and Wildlife Service, Washington, DC.

Schulin, R., A. Desaules, R. Webster and B. Von Steiger (eds) (1993). *Soil Monitoring: Early Detection and Surveying of Soil Contamination and Degradation*. Birkhäuser Verlag, Basel, Switzerland.

Schwab, G.O., R.K. Frevert, T.W. Edminster and K.K. Barnes (1966). *Soil and Water Conservation Engineering*, 2nd edn. John Wiley & Sons, New York.

Schwartz, S.I. and W.B. Pratt (1990). *Hazardous Waste from Small Quantity Generators – Strategies and Solutions for Business and Government*. Island Press, Washington, DC.

Schwing, R.C. and W.A. Albers, Jr (eds) (1980). *Societal Risk Assessment: How Safe is Safe Enough?* Plenum Press, New York.

Searle, C.E. (ed.) (1976). *Chemical Carcinogens*. ACS Monograph 173. American Chemical Society, Washington, DC.

Sebek, V. (ed.) (1990). Maritime transport, control and disposal of hazardous waste. *Marine Policy Int. J.* (Special Issue) **14**(3), May.

Sedman, R.M. (1989). The development of applied action levels for soil contact: a scenario for the exposure of humans to soil in a residential setting. *Environ. Health Persp.* **79**, 291–313.

Sharp, V.F. (1979). *Statistics for the Social Sciences*. Little, Brown & Co., Boston, Massachusetts.

Shrader-Frechette, K.S. (1985). *Risk Analysis and Scientific Method*. D. Reidel Publishing Co., Boston, Massachusetts.

Sims, R.C. and J.L. Sims (1986). Cleanup of contaminated soils. In: *Utilization, Treatment, and Disposal of Waste on Land*. Soil Science Society of America, Madison, Wisconsin, pp. 257–278.

Sincero, A.P. and G.A. Sincero (1996). *Environmental Engineering: A Design Approach*. Prentice Hall, Upper Saddle River, New Jersey.

Singhroy, V.H., D.D. Nebert and A.J. Johnson (eds) (1996). *Remote Sensing and GIS for Site Characterization: Applications and Standards.* ASTM Publication No. STP 1279, ASTM, Philadelphia, Pennsylvania.

Sitnig, M. (1985). *Handbook of Toxic and Hazardous Chemicals and Carcinogens.* Noyes Data Corp., Park Ridge, New Jersey.

Sittig, M. (1994). *World-Wide Limits for Toxic and Hazardous Chemicals in Air, Water and Soil.* Noyes Publications, Park Ridge, New Jersey.

Smith, A.H. (1987). Infant exposure assessment for breast milk dioxins and furans derived from waste incineration emissions. *Risk Anal.* **7**, 347–353.

Smith, L., J. Means and E. Barth (1995). *Recycling and Reuse of Industrial Wastes.* Battelle Press, Columbus, Ohio.

Smith, L.A. and R.E. Hinchee (1993). *In-Situ Thermal Technologies for Site Remediation.* Lewis Publishers/CRC Press, Boca Raton, Florida.

Spencer, E.Y. (1982). *Guide to the Chemicals Used in Crop Protection*, 7th edn. Research Institute, Agriculture Canada, Information Canada, Ottawa, Publ No. 1093.

Splitstone, D.E. (1991). How clean is clean . . . statistically? *Pollut. Engnr.* 90–96.

Sposito, G., C.S. LeVesque, J.P. LeClaire and N. Sensi (1984). Methodologies to predict the mobility and availability of hazardous metals in sludge-amended soils. California Water Resource Center, Contribution No. 189. University of California.

Starr, C. and C. Whipple (1980). Risks of risk decisions. *Science* **208**, 1114.

Starr, C., R. Rudman and C. Whipple (1976). Philosophical basis for risk analysis. *Ann. Rev. Energy* **1**, 629–662.

States, J.B., P.T. Hang, T.B. Schoemaker, L.W. Reed and E.B. Reed (1978). *A System Approach to Ecological Baseline Studies.* FQS/DBS-78/21. USFWS. Washington, DC.

Suess, M.J. and J.W. Huismans (eds) (1983). *Management of Hazardous Wastes: Policy Guidelines and Code of Practice.* WHO Regional Publication, European Series No. 14, Copenhagen. World Health Organization, Regional Office for Europe.

Talbot, E.O. and G.F. Craun (eds) (1995). *Introduction to Environmental Epidemiology.* Lewis Publishers/CRC Press, Boca Raton, Florida.

Tardiff, R.G. and J.V. Rodricks (eds) (1987). *Toxic Substances and Human Risk.* Plenum Press, New York, 445pp.

Tasca, J.J., M.F. Saunders and R.S. Prann (1989). *Terrestrial Food-Chain Model for Risk Assessment.* In Superfund '89: Proceedings 10th National Conference.

Tedder, D.W. and F.G. Pohland (eds) (1990). *Emerging Technologies in Hazardous Waste Management.* ACS Symposium Series 422, American Chemical Society, Washington, DC.

Tedder, D.W. and F.G. Pohland (eds) (1991). *Emerging Technologies in Hazardous Waste Management II.* ACS Symposium Series 468, American Chemical Society, Washington, DC.

Tedder, D.W. and F.G. Pohland (eds) (1993). *Emerging Technologies in Hazardous Waste Management III.* ACS Symposium Series 518, American Chemical Society, Washington, DC.

Tedder, D.W. and F.G. Pohland (eds) (1994). *Emerging Technologies in Hazardous Waste Management IV.* ACS Symposium Series 554, American Chemical Society, Washington, DC.

Tedder, D.W. and F.G. Pohland (eds) (1995). *Emerging Technologies in Hazardous Waste Management V.* ACS Symposium Series 607, American Chemical Society, Washington, DC.

Testa, S.M. (1994). *Geological Aspects of Hazardous Waste Management.* Lewis Publishers/CRC Press, Boca Raton, Florida.

Testa, S.M. and D.L. Winegardner (1990). *Restoration of Petroleum-Contaminated Aquifers.* Lewis Publishers/CRC Press, Boca Raton, Florida.

The Conservation Foundation (1985). *Risk Assessment and Risk Control.* The Conservation Foundation, Washington, DC.

Theiss, J.C. (1983). The ranking of chemicals for carcinogenic potency. *Regulatory Toxicol. Pharmacol.* **3**, 320–328.

Theodore, L., J.P. Reynolds and F.B. Taylor (1989). *Accident and Emergency Management.* Wiley-Interscience, New York.

The World Bank (1985). *Manual of Industrial Hazard Assessment Techniques.* Office of Environment and Scientific Affairs, Washington, DC.

The World Bank (1989). *Striking a Balance – The Environmental Challenge of Development.* IBRD/The World Bank, Washington, DC.

Thibodeaux, L.J. (1979). *Chemodynamics: Environmental Movement of Chemicals in Air, Water and Soil.* John Wiley & Sons, New York.

Thibodeaux, L.J. and S.T. Hwang (1982). Landfarming of petroleum wastes – modeling the air emission problem. *Environmental Progress* **1**, 42–46.

Thompson, S.K. (1992). *Sampling.* John Wiley & Sons, New York.

Tolba, M.K. (ed.) (1988). *Evolving Environmental Perceptions: From Stockholm to Nairobi.* United Nations Environment Programme, UNEP, Nairobi.

Tolba, M.K. (1990). The global agenda and the hazardous waste challenge. *Marine Policy Int. J.* **14**(3), May, 205–209.

Travis, C.T. and A.D. Arms (1988). Bioconcentration of organics in beef, milk, and vegetation. *Environ. Sci. Technol.* **22**, 271.

Travis, C.C. and H.A. Hattemer-Frey (1988). Determining an acceptable level of risk. *Environ. Sci. Technol.* **22**(8).

USBR (US Bureau of Reclamation) (1986). *Guidelines to Decision Analysis.* ACER Tech. Memo. No. 7, Denver, Colorado.

USEPA (US Environmental Protection Agency) (1982). *Test Methods for Evaluating Solid Waste: Physical/Chemical Methods,* 1st edn. SW-846. USEPA.

USEPA (US Environmental Protection Agency) (1983). *Hazardous Waste Land Treatment.* Revised edn. SW-8974. US EPA, Cincinnati, OH.

USEPA (US Environmental Protection Agency) (1984). *Approaches to Risk Assessment for Multiple Chemical Exposures.* US Environmental Protection Agency, Environmental Criteria and Assessment Office, Cincinnati, Ohio, EPA-600/9-84-008.

USEPA (US Environmental Protection Agency) (1984). Proposed guidelines for carcinogen, mutagenicity, and developmental toxicant risk assessment. *Federal Register* **49**, 46294–46331.

USEPA (US Environmental Protection Agency) (1985). *Chemical, Physical, and Biological Properties of Compounds Present at Hazardous Waste Sites.* Report Prepared by Clement Associates for the USEPA (September 1985).

USEPA (US Environmental Protection Agency) (1985). *Development of Statistical Distribution or Ranges of Standard Factors Used in Exposure Assessments.* US Environmental Protection Agency, Office of Health and Environmental Assessment, Washington, DC.

USEPA (US Environmental Protection Agency) (1985). *Practical Guide to Ground-Water Sampling.* Robert S. Kerr Environmental Research Lab., Office of Research and Development, USEPA, Ada, OK. EPA/600/2-85/104 (September 1985).

USEPA (US Environmental Protection Agency) (1985). *Rapid Assessment of Exposure to Particulate Emissions from Surface Contamination Sites.* EPA/600/8-85/002. NTIS PB85-192219. Office of Health and Environmental Assessment, Washington, DC.

USEPA (US Environmental Protection Agency) (1986). *Ecological Risk Assessment.* Hazard Evaluation Division Standard Evaluation Procedure. Washington, DC.

USEPA (US Environmental Protection Agency) (1986). Guidelines for carcinogen risk assessment. *Federal Register* **51**(185), 33992–34003, CFR 2984, 24 September 1986.

USEPA (US Environmental Protection Agency) (1986). *Methods for Assessing Exposure to Chemical Substances,* Vol. 8 – *Methods for Assessing Environmental Pathways of Food Contamination.* Exposure Evaluation Division, Office of Toxic Substances. EPA 560/5-85-008 (September 1986).

USEPA (US Environmental Protection Agency) (1986). *Registry of Toxic Effects of Chemical Substances.* Research Triangle Park, North Carolina.

USEPA (US Environmental Protection Agency) (1986). *Superfund Public Health Evaluation Manual.* EPA/540/1-86/060. Office of Emergency and Remedial Response, Washington, DC.

USEPA (US Environmental Protection Agency) (1986). *Superfund Risk Assessment Information Directory.* EPA/540/1-86/061. Office of Emergency and Remedial Response, Washington, DC.

USEPA (US Environmental Protection Agency) (1987). *Data Quality Objectives for Remedial Response Activities: Example Scenario.* EPA/540/G-87/004. USEPA, Washington, DC.

USEPA (US Environmental Protection Agency) (1987). *Handbook for Conducting Endangerment Assessments.* USEPA, Research Triangle Park, North Carolina.

USEPA (US Environmental Protection Agency) (1987). *Quality Assurance Program Plan.* EPA/ 600/X-87/241. Quality Assurance Management Staff, USEPA, Las Vegas, Nevada.
USEPA (US Environmental Protection Agency) (1987). *Technical Guidance for Hazard Analysis.* Washington, DC.
USEPA (US Environmental Protection Agency) (1988). *A Workbook of Screening Techniques for Assessing Impacts of Toxic Air Pollutants.* EPA-450/4-88-009. Office of Air Quality Planning and Standards. Research Triangle Park, North Carolina.
USEPA (US Environmental Protection Agency) (1988). *Estimating Toxicity of Industrial Chemicals to Aquatic Organisms Using Structure Activity Relationships.* Office of Toxic Substances, EPA/560/6-88/001.
USEPA (US Environmental Protection Agency) (1988). *Review of Ecological Risk Assessment Methods.* Office of Policy, Planning and Evaluation.
USEPA (US Environmental Protection Agency) (1988). *CERCLA Compliance with Other Laws Manual.* EPA/540/6-89/006. Office of Solid Waste and Emergency Response, Washington, DC.
USEPA (US Environmental Protection Agency) (1989). *CERCLA Compliance with Other Laws Manual: Part II – Clean Air Act and Other Environmental Statutes and State Requirements.* EPA/540/G-89/009. OSWER Directive 9234.1-02.
USEPA (US Environmental Protection Agency) (1989). *Ecological Assessments of Hazardous Waste Sites: A Field and Laboratory Reference Document.* Office of Research and Development – Corvallis Environmental Research Laboratory, Oregon, EPA/600/3-89/013.
USEPA (US Environmental Protection Agency) (1989). *Application of Air Pathway Analyses for Superfund Activities. Air/Superfund National Technical Guidance Study Series. Procedures for Conducting Air Pathway Analyses for Superfund Applications, Volume 1.* EPA-450/1-89-001. Interim Final. Office of Air Quality Planning and Standards, Research Triangle Park, North Carolina.
USEPA (US Environmental Protection Agency) (1989). *Estimation of Air Emissions from Cleanup Activities at Superfund Sites. Air/Superfund National Technical Guidance Study Series, Volume III.* EPA-450/1-89-003. Interim Final. Office of Air Quality Planning and Standards, Research Triangle Park, North Carolina.
USEPA (US Environmental Protection Agency) (1989). *Procedures for Conducting Air Pathway Analyses for Superfund Applications. Volume IV – Procedures for Dispersion Modeling and Air Monitoring for Superfund Air Pathway Analyses. Air/Superfund National Technical Guidance Study Series.* EPA-450/1-89-004. Interim Final. Office of Air Quality Planning and Standards, Research Triangle Park, North Carolina.
USEPA (US Environmental Protection Agency) (1989). *Review and Evaluation of Area Source Dispersion Algorithms for Emission Sources at Superfund Sites.* EPA-450/4-89-020. Office of Air Quality Planning and Standards, Research Triangle Park, North Carolina.
USEPA (US Environmental Protection Agency) (1989). *Estimating Air Emissions from Petroleum UST Cleanups.* Office of Underground Storage Tanks, Washington, DC.
USEPA (US Environmental Protection Agency) (1989). *Exposure Assessment Methods Handbook.* Office of Health and Environmental Assessment, USEPA, Ohio.
USEPA (US Environmental Protection Agency) (1989). *Methods for Evaluating the Attainment of Cleanup Standards, Vol. I – Soils and Solid Media.* EPA/230/2-89/042. Office of Policy, Planning and Evaluation, Washington, DC.
USEPA (US Environmental Protection Agency) (1990). *Emission Factors for Superfund Remediation Technologies.* Draft. Office of Air Quality Planning and Standards, RTP, NC.
USEPA (US Environmental Protection Agency) (1990). *Estimation of Baseline Air Emissions at Superfund Sites. Air/Superfund National Technical Guidance Study Series. Procedures for Conducting Air Pathway Analyses for Superfund Applications. Volume II.* EPA-450/1-89-002a. Office of Air Quality Planning and Standards, Research Triangle Park, North Carolina.
USEPA (US Environmental Protection Agency) (1990). Hazard ranking system. *Federal Register* **55**(241), December, 51532–51666.
USEPA (US Environmental Protection Agency) (1990). *Air/Superfund National Technical Guidance Study Series. Development of Example Procedures for Evaluating the Air Impacts of*

Soil Excavation Associated with Superfund Remedial Actions. EPA-450/4-90-014. Office of Air Quality Planning and Standards, Research Triangle Park, North Carolina.

USEPA (US Environmental Protection Agency) (1991). *Conducting Remedial Investigations/ Feasibility Studies for CERCLA Municipal Landfill Sites*. Office of Emergency and Remedial Response, Washington, DC. EPA/540/P-91/001 (OSWER Directive 9355.3-11).

USEPA (US Environmental Protection Agency) (1991). *Guidance for Performing Site Inspections Under CERCLA* – Interim Version. Draft Publication, OSWER Directive 9345.1-06. Office of Emergency and Remedial Response, Washington, DC.

USEPA (US Environmental Protection Agency) (1991). *Risk Assessment Guidance for Superfund:* Vol. I – *Human Health Evaluation Manual (Part B, Development of Risk-based Preliminary Remediation Goals)*. PB92-963333, OSWER Directive: 9285.7-01B, Interim (October 1991). Office of Emergency and Remedial Response, Washington, DC.

USEPA (US Environmental Protection Agency) (1991). *Risk Assessment Guidance for Superfund:* Vol. I – *Human Health Evaluation Manual (Part C, Risk Evaluation of Remedial Alternatives)*. PB90-155581, OSWER Directive: 9285.01C, Interim (December 1991). Office of Emergency and Remedial Response, Washington, DC.

USEPA (US Environmental Protection Agency) (1991). *The Role of Baseline Risk Assessment in Superfund Remedy Selection Decisions*. Office of Solid Waste and Emergency Response, Washington, DC. OSWER Directive: 9355.0-30.

USEPA (US Environmental Protection Agency) (1992). *Framework for Ecological Risk Assessment*. EPA/630/R-92/001, February 1992, Washington, DC.

USEPA (US Environmental Protection Agency) (1992). *Guideline for Predictive Baseline Emissions Estimation Procedures for Superfund Sites*. In: *Air/Superfund National Technical Guidance Study Series*. Interim Final. Office of Health and Environmental Assessment. EPA 450/I-92-002.

USEPA (US Environmental Protection Agency) (1993). *National Ambient Air Quality Standard for Particulate Matter*. 40 CFR. Part 50.6.

USEPA (US Environmental Protection Agency) (1994). *Technical Background Document for Soil Screening Guidance*. EPA/540/R-94/102. Office of Emergency and Remedial Response, Washington, DC. PB95-9633530.

USEPA (US Environmental Protection Agency) (1995). *Vendor Information System for Innovative Treatment Technologies*. EPA-542-C-95-001, Technology Innovation Office, US EPA, Washington, DC.

USEPA–NWWA (1989). *Handbook of Suggested Practices for the Design and Installation of Ground Water Monitoring Wells*. National Water Well Association, Dublin, Ohio.

Van den Brink, W.J., R. Bosman and F. Arendt (eds) (1995). *Contaminated Soil '95, Volumes I & II*. Kluwer Academic Publishers, Dordrecht, The Netherlands.

Van Ryzin, J. (1980). Quantitative risk assessment. *J. Occ. Med.* **22**, 321–326.

Verschueren, K. (1983). *Handbook of Environmental Data on Organic Chemicals*, 2nd edn. Van Nostrand Reinhold, New York.

Vesilind, P.A., J.J. Peirce and R.F. Weiner (1994). *Environmental Engineering, 3rd edition*. Butterworth-Heinemann, Newton, Mass.

Vidic, R.D. and F.G. Pohland (eds) (1995). Innovative Technologies for Site Remediation and Hazardous Waste Management. *Proceedings of the ASCE National Conf., American Society of Civil Engineers, New York*.

Volpp, C. (1988). 'Is it safe or isn't it?': An overview of risk assessment. *Water Resource News* **4**(1). New Jersey Department of Environmental Protection, Division of Water Resources, New Jersey.

Wang, L.K. and M.H.S. Wang (eds) (1992). *Handbook of Industrial Waste Treatment*, Vol. 1. Mercel Dekker, Inc., New York.

Weast, R.C. (ed.) (1984). *Handbook of Chemistry and Physics*, 65th edn. CRC Press, Inc., Boca Raton, Florida.

Whipple, C. (1987). De Minimis Risk. *Contemporary Issues in Risk Analysis*, Vol. 2. Plenum Press, New York.

Whipple, W., Jr (1994). *New Perspectives in Water Supply*. Lewis Publishers, Boca Raton, Florida.

WHO (World Health Organization) (1983). *Management of Hazardous Waste*. WHO Regional European Series No. 14.

Whyte, A.V. and I. Burton (eds) (1980). *Environmental Risk Assessment*. SCOPE Report 15. John Wiley & Sons, New York.

Williams, J.R. (1975). Sediment-yield prediction with the universal equation using runoff energy factor. In: *Present and Prospective Technology for Predicting Sentiment Yields and Sources*. US Department of Agriculture. ARS-S-40.

Wilson, R. and E.A.C. Crouch (1987). Risk assessment and comparisons: an introduction. *Science* **236**, 267–270.

Wilson, L.G., L.G. Everett and S.J. Cullen (eds) (1995). *Handbook of Vadose Zone Characterization and Monitoring*. Lewis Publishers/CRC Press, Boca Raton, Florida.

Wilson, D.J. (1995). *Modeling of In Situ Techniques for Treatment of Contaminated Soils*. Technomic Publishing Co., Inc., Lancaster, Pennsylvania.

Wise, D.L. and D.J. Trantolo (eds) (1994). *Remediation of Hazardous Waste Contaminated Soils*. Marcel Dekker, New York.

Woodside, G. (1993). *Hazardous Materials and Hazardous Waste Management: A Technical Guide*. John Wiley & Sons, New York.

Worster, D. (1993). *The Wealth of Nature: Environmental History and the Ecological Imagination*. Oxford University Press, New York.

WPCF (Water Pollution Control Federation) (1990). *Hazardous Waste Site Remediation Management*. A Special Publication of the WPCF. Technical Practice Committee, Alexandria, Va.

Yakowitz, H. (1989). *Monitoring and Control of Transfrontier Movements of Hazardous Wastes: An International Overview*. Technical Paper W/0587M, OECD, Paris.

Yakowitz, H. (1990). Monitoring and control of transfrontier movements of hazardous wastes: an international overview. In: K.L. Zirm and J. Mayer (eds), *The Management of Hazardous Substances in the Environment*. Elsevier Applied Science, London, England, pp. 139–162.

Yang, J.T. and W.E. Bye (1979). *A Guidance for Protection of Ground Water Resources from the Effects of Accidental Spills of Hydrocarbons and other Hazardous Substances*. EPA-570/9-79-017, USEPA, Washington, DC.

Yin, S.C.L. (1988). *Modeling Groundwater Transport of Dissolved Gasoline*. Specialty Conference Proceedings, Joint CSCE–ASCE National Conference on Environmental Engineering, Vancouver, Canada, 13–15 July 1988, pp. 544–551.

Zirm, K.L. and J. Mayer (eds) (1990). *The Management of Hazardous Substances in the Environment*. Elsevier Applied Science, London, UK.

Zogg, H.A. (1987). *'Zurich' Hazard Analysis*. Zurich Insurance Group, Risk Engineering, Zurich, Switzerland.

Zoller, U. (ed.) (1994). *Groundwater Contamination and Control*. Marcel Dekker, New York.

Recommended scientific journals

American Journal of Industrial Medicine, J. Wiley & Sons, Chichester, UK.
Aquatic Conservation: Marine and Freshwater Ecosystems, J. Wiley & Sons, Chichester, UK.
Aquatic Toxicology, Elsevier Science BV, Amsterdam, The Netherlands.
Archives of Environmental Contamination and Toxicology, Springer-Verlag, New York.
Biodegradation, Kluwer Academic Publishers, Dordrecht, The Netherlands.
Clean Air (and Environmental Protection), National Society for Clean Air and Environmental Protection, Brighton, UK.
Clean Air, The Journal of the Clean Air Society of Australia and New Zealand, Eastwood, NSW, Australia.
Contaminant Hydrology, Elsevier Science, Amsterdam, The Netherlands.
Critical Reviews in Environmental Science and Technology, CRC Press, Inc., Boca Raton, Florida.
Critical Reviews in Toxicology, CRC Press, Inc., Boca Raton, Florida.
Ecological Applications: A Publication of the Ecological Society of America, Washington, DC.
Ecological Modelling: International Journal on Ecological Modelling and Systems Ecology, Elsevier Science BV, Amsterdam, The Netherlands.
Ecological Research, The Ecological Society of Japan, Blackwell Science (Australia) Pty. Ltd, Victoria, Australia.
Ecotoxicology and Environmental Safety, Academic Press, Inc., Orlando, Florida.
Environmental and Molecular Mutagenesis, J. Wiley & Sons, Chichester, UK.
Environmental Engineering Geoscience, AEG, Texas A&M University, College Station, Texas.
Environmental Geochemistry and Health, Chapman & Hall, London, UK.
Environmental Health Perspectives Supplements, National Institute of Environmental Health Sciences, Research Triangle Park, North Carolina.
Environmental Health Perspectives: Journal of the National Institute of Environmental Health Sciences, Research Triangle Park, North Carolina.
Environmental Impact Assessment Review, Elsevier Science Inc., New York.
Environmental Law and Management, J. Wiley & Sons, Chichester, UK.
Environmental Management, Springer-Verlag, New York.
Environmental Monitoring and Assessment: An International Journal, Kluwer Academic Publishers, Dordrecht, The Netherlands.
Environmental Research: A Journal of Environmental Medicine and the Environmental Sciences, Academic Press, San Diego, California.
Environmental Science & Technology, American Chemical Society, Washington, DC.
Environmental Toxicology and Chemistry: An International Journal, Pergamon, Elsevier Science Ltd, Oxford, UK.
Environmental Toxicology and Water Quality, J. Wiley & Sons, Chichester, UK.
EnvironMetrics, John Wiley & Sons, Northampton, UK.
Hazardous Waste & Hazardous Materials, Mary Ann Liebert, Inc., Publishers, Larchmont, New York.
Human & Experimental Toxicology: An International Journal, British Toxicology Society, Macmillan Press Ltd, Hampshire, UK.
Hydrological Processes: An International Journal, J. Wiley & Sons, Chichester, UK.
Indoor Environment, KARGER, Rothenfluh, Switzerland.
International Journal of Climatology, J. Wiley & Sons, Chichester, UK.
International Journal of Environment and Pollution, Inderscience Enterprises Ltd/UNESCO, Geneva, Switzerland.
International Journal of Environmental Health Research, CARFAX Publishing Company, Oxfordshire, UK.
Issues in Science and Technology, J. Wiley & Sons, Chichester, UK.
Journal of Applied Toxicology, John Wiley & Sons Ltd, Chichester, UK.

Journal of Arid Environments, Academic Press, London, UK.

Journal of Clean Technology and Environmental Sciences, Princeton Scientific Publishing Company, Inc., Princeton, New Jersey.

Journal of Environmental Engineering, American Society of Civil Engineers, ASCE, New York, NY.

Journal of Environmental Pathology, Toxicology, and Oncology, Begell House, Inc., New York.

Journal of Environmental Science and Health, Marcel Dekker, Inc., New York.

Journal of Environmental Science and Health, Part A: Environmental Science and Engineering and Toxic and Hazardous Substance Control, Marcel Dekker, Inc., Monticello, New York.

Journal of Environmental Science and Health, Part B: Pesticides, Food Contaminants, and Agricultural Wastes, Marcel Dekker, Inc., Monticello, New York.

Journal of Environmental Science and Health, Part C: Environmental Carcinogenesis & Ecotoxicology Reviews, Marcel Dekker, Inc., Monticello, New York.

Journal of Environmental Systems, Baywood Publishing Company, Inc., Amityville, New York.

Journal of Environmental Technology, Publications Division, Selper Ltd, London, UK.

Journal of Hazardous Materials, Elsevier, Amsterdam, The Netherlands.

Journal of Soil Contamination, Lewis Publishers/CRC Press, Boca Raton, Florida.

Journal of the Air & Waste Management Association, AWMA, Pittsburgh, Pennsylvania.

Journal of the Institution of Water and Environmental Management, IWEM, London, UK.

Journal of Toxicology and Environmental Health, Taylor & Francis, London, UK.

Land Degradation and Rehabilitation, J. Wiley & Sons, Chichester, UK.

Natural Toxins, J. Wiley & Sons, Chichester, UK.

Pesticide Science, J. Wiley & Sons, Chichester, UK.

Restoration Ecology: The Journal of the Society for Ecological Restoration, Blackwell Science, Inc., Cambridge, Massachusetts.

Risk Analysis: An International Journal, Plenum Press, New York.

The Environmental Professional, NAEP, Blackwell Science, Inc., Cambridge, Massachusetts.

The Environmentalist, Chapman & Hall, Hampshire, UK.

The Science of the Total Environment (An International Journal for Scientific Research into the Environment and its Relationship with Man), Elsevier Science BV, Amsterdam, The Netherlands.

Toxic Substance Mechanisms, Taylor & Francis, London, UK.

Toxicology and Industrial Health: An International Journal, Princeton Scientific Publishing Company, Inc., Princeton, New Jersey.

Toxicology Letters, Elsevier Science Ireland Ltd, Shannon, Ireland.

Toxicology, Elsevier Science BV, Amsterdam, The Netherlands.

Waste Management & Research, ISWA, Copenhagen, Denmark.

Waste Management, Pergamon/Elsevier Science Ltd, Oxford, UK.

Water, Air & Soil Pollution: An International Journal of Environmental Pollution, Kluwer Academic Publishers, Dordrecht, The Netherlands.

Water, Air & Soil Pollution: An International Journal of Environmental Pollution, Kluwer Academic Publishers, London, UK.

Appendix A

Selected abbreviations

ADD	average daily dose
ADI	acceptable daily intake
BCF	bioconcentration (or bioaccumulation) factor
BTEX	benzene, toluene, ethylbenzene, and xylene
CSM	conceputal site model
(C)SF	(cancer) slope factor
DDT	dichlorodiphenyl trichloroethane
DNAPL	dense non-aqueous phase liquid
DQO	data quality objective
FS	feasibility study
FSP	field sampling plan
GAC	granular activated carbon
GC/PID	gas chromatograph/photoionization detector
HSP	health and safety plan
IRPTC	International Register of Potentially Toxic Chemicals
ISV	*in situ* vitrification
LADD	lifetime average daily dose
LNAPL	light non-aqueous phase liquid
MDD	maximum daily dose
NAPL	non-aqueous phase liquid
ND	non-detect (for analytical results)
NFA	no further action
NFAD	no-further-action decision
OVA/GC	organic vapor analyzer/gas chromatograph
PAH	polyaromatic hydrocarbon
PCB	polychlorinated biphenyl
PQL	practical quantitation limit (see also, SQL)
QA/QC	quality assurance/quality control
RBCL	risk-based cleanup level
RfD	reference dose
RI	remedial investigation
RI/FS	remedial investigation/feasibility study
SAP	sampling and analysis plan
SQL	sample quantitation limits (see also, PQL)
SVE	soil vapor extraction
TCE	trichloroethylene (or trichloroethene)
TPH	total petroleum hydrocarbon

UCL (95%)	95% upper confidence level
UCR	unit cancer risk
UST	underground storage tank
VES	vapor extraction system
VOC	volatile organic compound/chemical

Appendix B

Glossary of selected terms

Absorbed dose   The amount of a chemical substance actually entering an exposed organism via the lungs (for inhalation exposures), the gastrointestinal tract (for ingestion exposures), and/or the skin (for dermal exposures). It represents the amount penetrating the exchange boundaries of the organism after contact.

Absorption   The transport of a substance through the outer boundary of a medium. Generally used to refer to the uptake of a chemical by a cell or an organism, including the flow into the bloodstream following exposure through the skin, lungs, and/or gastrointestinal tract.

Absorption factor   The percent or fraction of a chemical in contact with an organism that becomes absorbed into the receptor.

Acceptable daily intake (ADI)   An estimate of the maximum amount of a chemical (in mg/kg body weight/day) to which a potential receptor can be exposed to on a daily basis over an extended period of time – usually a lifetime – without suffering a deleterious effect, or without anticipating an adverse effect.

Acceptable risk   A risk level generally deemed by society to be acceptable or tolerable.

Activated carbon   A highly adsorbent form of carbon used to remove contaminants from fluidal emissions or discharges. It is a special form of carbon, often derived from charcoal and treated to make it capable of adsorbing and retaining certain chemical substances.

Activated carbon adsorption   A treatment technology based on the principle that certain organic constituents preferentially adsorb to organic carbon.

Adsorption   The removal of contaminants from a fluid stream by concentration of the constituents on to a solid material. It is the physical process of attracting and holding molecules of other chemical substances on the surface of a solid, usually by the formation of chemical bonds. A substance is said to be adsorbed if the concentration in the boundary region of a solid (e.g. soil) particle is greater than in the interior of the contiguous phase.

Air sparging   The process of blowing air through a liquid for mixing purposes, in order to strip volatile materials (i.e. VOCs) or to add oxygen. Usually refers to the highly controlled injection of air into a contaminant plume in the soil *saturated* zone. This consists of the injection of air below the water table to strip volatile contaminants from the saturated zone.

Air stripping   A remediation technique that involves the physical removal of dissolved-phase contamination from a water stream.

Aquifer   A geological formation, group of formation, or part of a formation which is capable of yielding significant and usable quantities of groundwater to wells and/or springs.

Arithmetic mean (also, Average)   A statistical measure of central tendency for data from a normal distribution, defined for a set of n values, by the sum of values divided by n:

$$X_{\mathrm{m}} = \frac{\sum_{i=1}^{n} X_i}{n}$$

Asphalt batching   Also referred to as *asphalt incorporation*, is a method for treating hydrocarbon-contaminated soils. It involves the incorporation of a petroleum-laden soils into hot asphalt mixes as a partial substitute for stone aggregate. This mixture can then be utilized for pavings.

Attenuation   Any decrease in the amount or concentration of a pollutant in an environmental matrix as it moves in time and space. It is the reduction or removal of contaminant constituents by a combination of physical, chemical, and/or biological factors acting upon the contaminated media.

Auger   A rotary drilling equipment used in soils or unconsolidated materials, that continuously removes cuttings from a borehole by mechanical means without the use of fluids.

Average concentration   A mathematical average of contaminant concentration(s) from more than one sample, typically represented by the arithmetic mean or the geometric mean for environmental samples.

Average daily dose (ADD)   The average dose calculated for the duration of receptor exposure, and used to estimate risks for chronic noncarcinogenic effects of environmental contaminants.

Background threshold level   The normal ambient environmental concentration of a chemical constituent. It may include both naturally-occurring concentrations and elevated levels resulting from non-site-related human activities.

Benchmark risk   A threshold level of risk, typically prescribed by regulations, above which corrective measures will almost certainly have to be implemented to mitigate the risks.

Bioaccumulation   The retention and concentration of a chemical by an organism. It is a build-up of a chemical in a living organism which occurs when the organism takes in more of the chemical than it can get rid of in the same length of time and stores the chemical in its tissue, etc.

Bioaugmentation   A process in which specially selected bacteria cultures that are predisposed to metabolize some target compound(s) are added to impacted media, along with the nutrient materials, to encourage degradation of the contaminants of concern.

Bioconcentration   The accumulation of a chemical substance in tissues of organisms (such as fish) to levels greater than levels in the surrounding media (such as water) for the organism's habitat; often used synonymously with bioaccumulation.

Bioconcentration factor (BCF)   A measure of the amount of selected chemical substances that accumulates in humans or in biota. It is the ratio of the concentration of a chemical substance in an organism at equilibrium to the concentration of the substance in surrounding environmental medium.

Biodegradable   Capable of being metabolized by a biologic process or an organism.

Biodegradation   Decomposition of a substance into simpler substances by the action of microorganisms, usually in soil. It may or may not detoxify the material which is decomposed.

Bioremediation   Also called *biorestoration*, is a viable and cost-effective remediation technique for treating a wide variety of contaminants (such as petroleum and aromatic hydrocarbons, chlorinated solvents, and pesticides). It relies on microorganisms to transform hazardous compounds found in environmental matrices into innocuous or less toxic metabolic products. *Natural in-situ bioremediation* involves the attenuation of contaminants by indigenous (native) microorganisms without any manipulation.

Biostimulation   A process whereby the addition of selected amounts of nutrient materials stimulate or encourage the growth of the indigenous bacteria in soil, resulting in the degradation of some target contaminant(s).

Biota   All living organisms which are found within a prescribed volume or space.

Borehole   A hole drilled into the earth, and into which a well casing or screen can be installed to construct a well.

Cancer slope factor (SF)   (also *Cancer potency factor*) Health effect information factor commonly used to evaluate health hazard potentials for carcinogens. It is a plausible upper-bound estimate of the probability of a response per unit intake of a chemical over a lifetime. That is, it is used to estimate an upper-bound probability of an individual developing cancer as a result of a lifetime of exposure to a particular level of a carcinogen.

Carcinogen   A chemical or substance capable of producing cancer in living organisms.

Carcinogenic   Capable of causing, and tending to produce or incite cancer in living organisms.

Cleanup   Actions taken to abate the situation involving the release or threat of release of contaminants that could potentially affect human health and/or the environment. This typically involves a process to remove or attenuate contamination levels, in order to restore the impacted media to an 'acceptable' or useable condition.

Cleanup level   The contaminant concentration goal of a remedial action, i.e. the concentration of media contaminant level to be attained through a remedial action.

Closure   All activities involved in taking a hazardous waste facility out of service and securing it for the duration required by applicable regulations and laws. Site closures typically follow the implementation of appropriate site restoration programs, with monitoring usually becoming part of the post-closure site activities.

Confidence interval (CI)   A statistical parameter used to specify a range and the probability that an uncertain quantity falls within this range.

Confidence limits, 95 percent (95% CL)   The limits of the range of values within which a single estimation will be included 95% of the time. For large sample sizes (i.e. $n > 30$),

$$95\%\mathrm{CL} = X_{\mathrm{m}} \pm \frac{1.96\sigma}{n^{0.5}}$$

where CL is the confidence level, and σ is the estimate of the standard deviation of the mean (X_{m}). For a limited number of samples $(n \leqslant 30)$, a confidence limit or confidence interval may be estimated from

$$\mathrm{CL} = X_{\mathrm{m}} \pm \frac{ts}{n^{0.5}}$$

where t is the value of the Student's t distribution (refer to standard statistical texts) for the desired confidence level and degrees of freedom, $(n-1)$.

Containment   Refers to systems used to prevent (or significantly reduce) the further spread of contamination. Such systems may consist of pumping (and/or injection) wells, and cut-off walls or covers designed and placed at strategic locations.

Contaminant   Any physical, chemical, biological, or radiological material that can potentially have adverse impacts on environmental media, or that can adversely impact public health and the environment. It represents any undesirable substance that is not naturally-occurring and therefore not normally found in the environmental media of concern.

Contaminant migration   The movement of a contaminant from its source through other matrices/media such as air, water, or soil. A *contaminant migration pathway* is the path taken by the contaminants as they travel from the contaminated site through various environmental media.

Contaminant plume   A body of contaminated groundwater or vapor originating from a specific source and spreading out due to influences of such factors as local groundwater conditions or soil vapor flow patterns. It represents the volume of groundwater or vapor containing the contaminants released from a pollution source.

Contaminant release   The ability of a contaminant to enter into other environmental media/ matrices (e.g. air, water or soil) from its source(s) of origin.

Corrective action   Action taken to correct a problem situation, such as remediation of chemical contamination in soil and groundwater.

Cost-effective alternative   The *most cost-effective alternative* is the lowest cost alternative that is technologically feasible and reliable, and which effectively mitigates and minimizes environmental damage. It generally provides adequate protection of public health, welfare, and/or the environment.

Data quality objectives (DQOs)   Qualitative and quantitative statements developed by analysts to specify the quality of data that, at a minimum, is needed and expected from a particular data collection activity (or site characterization activity). This is determined based on the end use of the data to be collected.

Decision framework   Management tool designed to facilitate rational decision-making on environmental contamination problems.

Degradation   The physical, chemical or biological breakdown of a complex compound into simpler compounds and byproducts.

Dense non-aqueous phase liquids (DNAPLs)   Organic liquids, composed of one or more contaminants that are more dense than water, often coalescing in an immiscible layer at the bottom of a saturated geologic unit (e.g. chlorinated solvents).

Dermal exposure   Exposure of an organism or receptor through skin absorption.

Detection limit   The minimum concentration or weight of analyte that can be detected by a single measurement with a known confidence level. *Instrument detection limit (IDL)* represents the lowest amount that can be distinguished from the normal 'noise' of an analytical instrument (i.e. the smallest amount of a chemical detectable by an analytical instrument under ideal conditions). *Method detection limit (MDL)* represents the lowest amount that can be distinguished from the normal 'noise' of an analytical method (i.e. the smallest amount of a chemical detectable by a prescribed or specified method of analysis).

Diffusion   The migration of molecules, atoms, or ions from one fluid to another in a direction tending to equalize concentrations.

Dissolved product   The water-soluble fuel components of contaminant releases.

Dose   The amount of a chemical taken in by potential receptors on exposure. It is a measure of the amount of the substance received by the receptor, whether human or animal, as a result of exposure, expressed as an amount of exposure (in mg) per unit body weight of the receptor (in kg).

Dose–response   The quantitative relationship between the dose of a chemical and an effect caused by exposure to such substance.

Ecosystem   The interacting system of a biological community and its abiotic (i.e. non-living) environment.

Endangerment assessment   A site-specific risk assessment of the actual or potential danger to human health and welfare and also the environment, from the release of hazardous chemicals into various environmental media.

Endpoint (toxic)   A biological effect used as index of the impacts of a chemical on an organism.

Environmental fate   The 'destination' of a chemical after release or escape into the environment, and following transport through various environmental compartments. It is the movement of a chemical through the environment by transport in air, water, and soil culminating in exposures to living organisms. It represents the disposition of a material in various environmental compartments (e.g. soil, sediment, water, air, biota) as a result of transport, transformation, and degradation.

Exposure   The situation of receiving a dose of a substance, or coming in contact with a hazard. It represents the contact of an organism with a chemical or physical agent available at the exchange boundary (e.g. lungs, gut, skin) during a specified time period.

Exposure assessment   The qualitative or quantitative estimation, or the measurement, of the dose or amount of a chemical to which potential receptors have been exposed, or could potentially be exposed to. It comprises of the determination of the magnitude, frequency, duration, route and extent of exposure (to the chemicals or hazards of potential concern).

Exposure conditions   Factors (such as location, time, etc.) that may have significant effects on an exposed population's response to a hazard situation.

Exposure duration   The length of time that a potential receptor is exposed to the contaminants of concern in a defined exposure scenario.

Exposure frequency   The number of times (per year or per event) that a potential receptor would be exposed to site contaminants in a defined exposure scenario.

Exposure parameters   Variables used in the calculation of intake (e.g. exposure duration, inhalation rate, average body weight).

Exposure pathway   The course a chemical or physical agent takes from a source to an exposed population or organism. It describes a unique mechanism by which an individual or population is exposed to chemicals or physical agents at or originating from a contaminated site.

Exposure point   A location of potential contact between an organism and a chemical or physical agent.

Exposure route   The avenue by which an organism contacts a chemical, such as inhalation, ingestion, and dermal contact.

Exposure scenario A set of conditions or assumptions about sources, exposure pathways, concentrations of chemicals, and potential receptors that aids in the evaluation and quantification of exposure in a given situation.

Feasibility study (FS) The analysis and selection of alternative remedial or corrective actions for hazardous waste or contaminated sites. The process identifies and evaluates remedial alternatives by utilizing a variety of appropriate environmental, engineering, and economic criteria.

Field sampling plan (FSP) A documentation that defines in detail the sampling and data gathering activities to be used in the investigation of a potentially contaminated site.

Free product Chemical constituent that floats on groundwater, or that remains 'unadulterated' in a contaminant pool in the environment.

Fugitive dust Atmospheric dust arising from disturbances of particulate matter exposed to the air. Fugitive dust emissions consist of the release of chemicals from contaminated surface soil into the air, attached to dust particles.

Geometric mean A statistical measure of the central tendency for data from a positively skewed distribution (lognormal), given by:

$$X_{gm} = [(X_1)(X_2)(X_3)...(X_n)]^{1/n}$$

or,

$$X_{gm} = antilog \left\{ \frac{\sum_{i=1}^{n} \log X_i}{n} \right\}$$

Groundwater Water beneath the ground surface. It represents underground waters, whether present in a well-defined aquifer, or present temporarily in the vadose (unsaturated soil) zone.

Hazard The inherent adverse effect that a chemical or other object poses. It is that innate character which has the potential for creating adverse and/or undesirable consequences. It defines the chance that a particular substance will have an adverse effect on human health or the environment in a particular set of circumstances which creates an exposure to that substance.

Hazard assessment The evaluation of system performance and associated consequences over a range of operating and/or failure conditions. It involves gathering and evaluating data on types of injury or consequences that may be produced by a hazardous situation or substance.

Hazard identification The systematic identification of potential accidents, upset conditions, etc. It is the recognition that a hazard exists and the definition of its characteristics. The process involves determining whether exposure to an agent can cause an increase in the incidence of a particular adverse health effect in receptors of interest.

Hazard index (HI) The sum of several hazard quotients for multiple substances and/or multiple exposure pathways.

Hazard quotient (HQ) The ratio of a single substance exposure level for a specified time period to a reference dose of that substance derived from a similar exposure period.

Hazardous substance Any substance that can cause harm to human health or the environment whenever excessive exposure occurs.

Hazardous waste Wastes that are ignitable, explosive, corrosive, reactive, toxic, radioactive, pathological or have some other property that produces substantial risk to life. It is that byproduct which has the potential of causing detrimental effects on human health and/or environment if not managed efficiently.

'Hot spot' Term used to denote zones where contaminants are present at much higher concentrations than the immediate surrounding areas. It represents a relatively small area which is highly contaminated within a study area.

Human health risk The likelihood (or probability) that a given exposure or series of exposures to a hazardous substance will cause adverse health impacts on individual receptors experiencing the exposures.

Hydrophilic Having greater affinity for water, or water-loving. Hydrophilic compounds tend to become dissolved in water.

Hydrophobic Tending not to combine with water, or less affinity for water. Hydrophobic compounds tend to avoid being dissolved in water and are more attracted to non-polar liquids (e.g. oils) or solids.

Immunoassay A site investigation method used to determine whether or not contaminants are present at a project site, based on their ability to bind to antibodies produced by a living organism in response to the contaminant.

Incineration A thermal treatment/degradation process by which contaminated materials are exposed to excessive heat in a variety of incinerator types. The incineration process typically involves the thermal destruction of contaminants by burning under controlled conditions. Depending on the intensity of the heat, the contaminants of concern are volatilized and/or destroyed during the incineration process.

Individual excess lifetime cancer risk An upper-bound estimate of the increased cancer risk, expressed as a probability, that an individual receptor could expect from exposure over a lifetime.

Ingestion An exposure type whereby chemical substances enter the body through the mouth, and into the gastrointestinal system.

Inhalation The intake of a substance by receptors through the respiratory tract system.

In situ In-place, i.e. within the contaminated material or matrix itself.

In situ vitrification The heating of the subsurface to extremely high temperatures in order to destroy organic contaminants and to 'entomb' inorganic constituents. Upon cooling, the treated materials solidify, incorporating inorganic contaminants and ash.

Intake The amount of material inhaled, ingested or dermally absorbed during a specified time period. It is a measure of exposure, expressed in mg/kg-day.

Interim action A preliminary action that initiates remediation of a contaminated site but may also constitute part of the final remedy.

Investigation-generated wastes (IGWs) Wastes generated in the process of collecting samples during a remedial investigation or site characterization activity. Such wastes must be handled according to all relevant and applicable regulatory requirements. The wastes may include soil, groundwater, used personal protective equipment, decontamination fluids, and disposable sampling equipment.

K_d *(Soil/water partition coefficient)* Provides a soil- or sediment-specific measure of the extent of chemical partitioning between soil or sediment and water, unadjusted for the dependence on organic carbon.

K_{oc} *(Organic carbon adsorption coefficient)* Provides a measure of the extent of chemical partitioning between organic carbon and water at equilibrium. It is a measure that indicates the extent to which a compound will sorb to the solid organic content of geologic media in the subsurface. It is computed as the ratio of the amount of chemical sorbed per unit weight of organic carbon in the soil to the concentration of the chemical in solution at equilibrium.

K_{ow} *(Octanol/water partition coefficient)* Provides a measure of the extent of chemical partitioning between water and octanol at equilibrium. It is a measure that indicates the extent to which a compound is attracted to an organic phase (for which octanol is a proxy) and hence the compound's tendency to sorb to subsurface materials. It is computed by dividing the amount that will dissolve in octanol by the amount that will dissolve in water. The greater the value, the greater the tendency to sorb in the subsurface.

K_w *(Water/air partition coefficient)* Provides a measure of the distribution of a chemical between water and air at equilibrium.

Landfarming The application of biodegradable organic wastes on to a land surface and their incorporation into the surface soil so that they degrade more readily.

Landfill A controlled site for the disposal of wastes on land, generally operated in accordance with regulatory safety and environmental compliance requirements.

Leachate Aqueous, often-contaminated, liquid generated when water percolates or trickles through waste materials or contaminated sites and collects components of those wastes. Leaching usually occurs at landfills as a result of infiltration of rainwater or snowmelt, and may result in hazardous chemicals entering soils, surface water, or groundwater.

Lifetime average daily dose (LADD) The exposure, expressed as mass of a substance contacted and absorbed per unit body weight per unit time, averaged over a lifetime. It is usually used to calculate carcinogenic risks; it takes into account the fact that, whereas carcinogenic risks are determined with an assumption of lifetime exposure, actual exposures may be for a shorter period of time.

Light non-aqueous phase liquids (LNAPLs) Organic fluids that are lighter than water, capable of forming an immiscible layer that floats on the water table; also referred to as 'floaters' (e.g. gasoline and fuel oil).

Matrix (or Medium) The predominant material comprising the environmental sample being investigated (e.g. soils, water, air).

Maximum daily dose (MDD) The maximum dose calculated for the duration of receptor exposure, and used to estimate risks for subchronic or acute noncarcinogenic effects of environmental contaminants.

Microbe A microscopic or ultramicroscopic organism (e.g. bacterium or virus).

Mitigation The process of reducing or alleviating a problem situation.

Modeling The use of mathematical equations to simulate and predict real events and processes.

Monitoring The measurement of concentrations of chemicals in environmental media or in tissues of humans and other biological receptors/organisms.

Non-aqueous phase liquid (NAPL) Organic compounds in the liquid phase that are not completely miscible in water.

Nonparametric statistics Statistical techniques whose application is independent of the actual distribution of the underlying population from which the data were collected.

Nonthreshold chemical Also called *zero threshold chemical*, refers to a substance which is known, suspected, or assumed to potentially cause some adverse response at any dose above zero.

Off-site Areas outside the boundaries or limits of a presumed contaminated site.

On-site The boundaries or limits of a presumed contaminated site.

Organic carbon content of soils or sediments (%) This reflects the amount of organic matter present, and generally correlates with the tendency of chemicals to accumulate in the soil or sediment. The accumulation of chemicals in soils or sediments is frequently the result of adsorption on to organic matter. Thus, in general, the higher the organic carbon content of the soil or sediment, the more a contaminant will be adsorbed to the soil particles, rather than be dissolved in the water or gases permeating the soil or sediment.

Partition coefficient A term used to describe the relative amount of a substance partitioned between two different phases, such as a solid and a liquid. It is the ratio of the chemical's concentration in one phase to its concentration in the other phase.

Partitioning A chemical equilibrium condition in which a chemical's concentration is apportioned between two different phases according to the partition coefficient.

Pathway Any specific route which environmental contaminants take in order to travel away from the source and to reach potential receptors or individuals.

Plume A zone containing predominantly dissolved (or vapor phase) contaminants and sorbed contaminants in equilibrium with the dissolved (or vapor phase) contaminants. A plume usually will originate from the contaminant source areas.

Preliminary assessment (PA) A survey and evaluation whereby sites are characterized with respect to their potential to release significant amounts of contaminants into the environment.

Preliminary site appraisal   Process used for quick assessment of a site's potential to adversely affect the environment and/or public health.

Probability   The likelihood of an event occurring.

Proxy concentration   Assigned contaminant concentration value for situations where sample data may not be available, or when it is impossible to quantify accurately.

Qualitative   Description of a situation without numerical specifications.

Quality assurance (QA)   A system of activities designed to assure that the quality control system is performing adequately. It consists of the management of investigations data to assure that they meet the data quality objectives.

Quality control (QC)   A system of specific efforts designed to test and control the quality of data obtained in an investigation. It consists of the management of activities involved in the collection and analysis of data to assure they meet the data quality objectives. It is the system of activities required to provide information as to whether the quality assurance system is performing adequately.

Quantitation limit (QL)   The lowest level at which a chemical can be accurately and reproducibly quantitated. It usually is equal to the instrument detection limit (IDL) multiplied by a factor of 3 to 5, but varies for different chemicals and different samples.

Quantitative   Description of a situation that is presented in exact numerical terms.

Receptor   Members of a potentially exposed population, such as persons or organisms that are potentially exposed to concentrations of a particular chemical compound.

Reference dose (RfD)   The maximum amount of a chemical that the human body can absorb without experiencing chronic health effects, expressed in mg of chemical per kg body weight per day. It is the estimate of lifetime daily exposure of a noncarcinogenic substance for the general human population (including sensitive receptors) which appears to be without an appreciable risk of deleterious effects, consistent with the threshold concept.

Remedial action   Those actions consistent with a permanent remedy in the event of a release of a hazardous substance into the environment, meant to prevent or minimize the effects of such releases so that they do not migrate farther to cause substantial danger to present or future public health or welfare or the environment.

Remedial action objective   Cleanup objectives that specify the level of cleanup, area of cleanup (or area of attainment), and the time required to achieve cleanup (i.e. the restoration timeframe).

Remedial alternative   An action considered in a feasibility study, that is intended to reduce or eliminate significant risks to human health and/or the environment at a contaminated site.

Remedial investigation (RI)   The field investigations of hazardous waste sites to determine pathways, nature, and extent of contamination, as well as to identify preliminary alternative remedial actions. It addresses data collection and site characterization to identify and assess threats or potential threats to human health and the environment posed by a site.

Remediation   The process of cleaning up a potentially contaminated site, in order to prevent or minimize the potential release and migration of hazardous substances from the impacted media that could cause adverse impacts to present or future public health and welfare, or the environment.

Removal action   An action that is implemented to address a direct threat to human health or the environment.

Representative sample   A sample that is assumed *not* to be significantly different than the population of samples available.

Residual risk   The risk of adverse consequences that remains after corrective actions have been implemented.

Response (toxic)   The reaction of a body or organ to a chemical substance or other physical, chemical, or biological agent.

Restoration time-frame   Time required to achieve requisite cleanup levels or site restoration goals.

Retardation coefficient A measure of how quickly a contaminant moves through the ground compared to water. It is computed as the ratio of the total contaminant mass in a unit aquifer volume to the contaminant mass in solution.

Risk The probability or likelihood of an adverse consequence from a hazardous situation or hazard, or the potential for the realization of undesirable adverse consequences from impending events. It is a measure of the probability and severity of an adverse effect to health, property, or the environment.

Risk acceptance The willingness of an individual, group, or society to accept a specific level of risk in order to obtain some gain or benefit.

Risk appraisal The assessment of whether existing or potential biologic receptors are presently, or may in the future be, at risk of adverse effects as a result of exposures to contaminants originating at a contaminated site.

Risk assessment A methodology that combines exposure assessment with health and environmental effects data to estimate risks to human or environmental target organisms which results from exposure to contaminants.

Risk-based concentration A contaminant concentration determined from an evaluation of the compound's overall risk to human health upon exposure.

Risk control The process to manage risks associated with a hazard situation. It may involve the implementation, enforcement, and re-evaluation of the effectiveness of corrective measures from time to time.

Risk decision The process used for making complex public policy decisions relating to the control of risks associated with hazardous situations.

Risk determination The evaluation of the environmental and health impacts of contaminant releases.

Risk estimate A description of the probability that a potential receptor exposed to a specified dose of a chemical will develop an adverse response.

Risk estimation The process of quantifying the probability and consequence values for a hazard situation. It is the process used to determine the extent and probability of adverse effects of the hazards identified, and to produce a measure of the level of health, property, or environmental risks being assessed.

Risk evaluation The complex process of developing acceptable levels of risk to individuals or society. It is the stage at which values and judgments enter into the decision process.

Risk group A real or hypothetical exposure group composed of the general or specific population groups.

Risk management The steps and processes taken to reduce, abate or eliminate the risk that has been revealed by a risk assessment. It is an activity concerned with decisions about whether an assessed risk is sufficiently high to present a public health concern, and about the appropriate means for controlling the risks judged to be significant.

Risk perception The magnitude of the risk as it is perceived by an individual or population. It consists of the measured risk and the pre-conceptions of the observer.

Risk reduction The action of lowering the probability of occurrence and/or the value of a risk consequence, thereby reducing the magnitude of the risk.

Sample blank Blanks are samples considered to be the same as the environmental samples of interest except with regards to one factor whose influence on the samples is being evaluated. Sample blanks are used to ensure that contaminant concentrations actually reflect site conditions, and are not artefacts of the sampling and sample handling processes. The blank consists of laboratory distilled, deionized water that accompanies the empty sample bottles to the field as well as the samples returning to the laboratory, where it is not opened until both blank and the actual site samples are ready to be analyzed.

Sample duplication Two samples taken from the same source at the same time and analyzed under identical conditions.

Sample quantitation limit (SQL) Also called *practical quantitation limit (PQL)*. It is the lowest level that can be reliably achieved within specified limits of precision and accuracy during routine laboratory operating conditions. It represents a detection limit that has been corrected for sample characteristics, sample preparation, and analytical adjustments such as dilution. Typically, the PQL or SQL will be about 5 to 10 times the chemical-specific detection limit.

Sampling and analysis plan (SAP)   Documentation that consists of a quality assurance project plan (QAPP) and a field sampling plan (FSP).

Saturated zone   An underground geologic formation in which the pore spaces or interstitial spaces in the formation are filled with water under pressure, equal to or greater than atmospheric pressure.

Sediment   Soil that is normally covered with water. It generally is considered to provide a direct exposure pathway to aquatic life.

Sensitive receptor   Individual in a population who is particularly susceptible to health impacts due to exposure to a chemical pollutant.

Site assessment   Process used to identify toxic substances that may be present at a site and to present site-specific characteristics that influence the migration of contaminants.

Site categorization   A classification of sites to reflect the uniqueness of each site.

Site characterization   A process that attempts to identify the types and sources of contaminants present at a site and the site's hydrogeologic characteristics.

Site cleanup   The decontamination of a site initiated as a result of the discovery of contamination at a site or property.

Site mitigation   The process of cleaning up a contaminated site in order to return it to an environmentally acceptable state.

Soil gas   The vapor or gas found in the unsaturated soil zone.

Soil vapor extraction (SVE)   Also known as *in situ soil venting, subsurface venting, vacuum extraction*, or *in situ soil stripping*, is a technique that uses soil aeration to treat subsurface zones of VOC contamination in soils.

Solubility   A measure of the ability of a substance to dissolve in a fluid.

Sorption   The processes that remove solutes from the fluid phase and concentrate them on the solid phase of a medium.

Source area   Sections at a contaminated site containing contamination that remains in place. The source area may stretch beyond the original contaminant spill site; included in the committee's definition of source area are regions along the contaminant flow path where contaminants are present in precipitated or nonaqueous-phase liquid form.

Stabilization   The conversion of a substance into a form that will not readily change its physical or chemical characteristics.

Standard   A general term used to describe legally established values above which regulatory action will be required.

Standard deviation   The most widely used statistical measure to describe the dispersion of a data set, defined for a set of *n* values as follows:

$$s = \left[\frac{\sum_{i=1}^{n} (X_i - X_m)^2}{(n-1)} \right]^{0.5}$$

where X_m is the arithmetic mean for the data set of *n* values.

Surfactant   A surface active chemical agent, usually made up of phosphates, used in detergents to produce lathering. It bonds with oils and other immiscible compounds to aid their transport in water.

Threshold   The lowest dose or exposure of a chemical at which a specified measurable effect is observed and below which such effect is not observed. *Threshold dose* is the minimum exposure dose of a chemical that will evoke a stipulated toxicological response. *Toxicological threshold* refers to the concentration at which a compound begins to exhibit toxic effects.

Threshold limit   A chemical concentration above which adverse health and/or environmental effects may occur.

Toxic   Harmful, or deleterious with respect to the effects produced by exposure to a chemical substance.

Toxicity The harmful effects produced by a chemical substance. It is the quality or degree of being poisonous or harmful to human or ecological receptors. It represents the property of a substance to cause any adverse physiological effects (on living organisms).

Toxicity assessment Evaluation of the toxicity of a chemical based on all available human and animal data. It is the characterization of the toxicological properties and effects of a chemical substance, with special emphasis on the establishment of dose–response characteristics.

Toxic substance Any material or mixture that is capable of causing an unreasonable threat to human health or the environment.

Treatment A change in the composition or concentration of a waste substance so as to make it less hazardous, or to make it acceptable at disposal and re-use facilities. It involves the application of technological process(es) to a contaminant or waste in order to render it non-hazardous or less hazardous or more suitable for resource recovery.

Trip blank A trip blank is transported just like actual samples, but does not contain the chemicals to be analyzed. The purpose of this blank is to evaluate the possibility that a chemical could seep into samples (to adulterate them) during transportation to the laboratory.

Uncertainty The lack of confidence in the estimate of a variable's magnitude or probability of occurrence.

Underground storage tank (UST) A tank fully or partially located below the ground surface, that is designed to hold gasoline or other petroleum products, or indeed other chemical products.

Unit cancer risk (UCR) The excess lifetime risk of cancer due to a continuous lifetime exposure/dose of one unit of carcinogenic chemical concentration (caused by one unit of exposure in the low exposure region).

Upper-bound estimate The estimate not likely to be lower than the true (risk) value.

Upper confidence limit, 95% (95% UCL) The upper limit on a normal distribution curve below which the observed mean of a data set will occur 95% of the time. This is also equivalent to stating that there is at most a 5% chance of the true mean being greater than the observed value.

Vadose zone Also called the *unsaturated soil zone*, is the zone between the ground surface and the top of the groundwater table.

Volatile organic compound (VOC) Any organic compound that has a great tendency to vaporize, and is susceptible to atmospheric photochemical reactions. It volatilizes (evaporates) relatively easily when exposed to air.

Volatility A measure of the tendency of a compound to vaporize from the liquid state.

Volatilization The transfer of a chemical from the liquid into the gaseous phase.

Water table The top of the saturated zone where confined groundwater is under atmospheric pressure.

Appendix C

Important elements and requirements of a site characterization activity program

Workplans are normally required to specify the administrative and logistic requirements of site investigation activities. As part of any site characterization program carried out for a potentially contaminated site problem, a carefully executed workplan is developed to guide all relevant decisions. A typical site characterization workplan will consist of the following elements:

- How site mapping will be performed (including survey limits, scale of site plan to be produced, horizontal and vertical control, and significant site features).
- Number of individuals to be involved in each field sampling task, and estimates of the duration of work.
- Identification of soil boring and test pit locations on a map to be provided in a detailed workplan.
- Number of samples to be obtained in the field (including blanks and duplicates), and the sampling locations (illustrated on maps to be included in a detailed workplan).
- An elaboration of how investigation-generated wastes will be handled.
- List of field and laboratory analyses to be performed.
- A general discussion of data quality objectives.
- Identification of pilot or bench-scale studies that will be performed, where necessary, in relationship to recommendations for remedial technologies screening, risk management strategies, and/or site stabilization processes.
- A discussion of health and safety plans required for the site investigation or site restoration activities, as well as that necessary to protect populations in the general vicinity of the site.

The major components and elements required of most site characterization activity programs are elaborated below.

C.1   Sampling and Analysis Considerations for the Investigation of Contaminated Sites

Sampling and analysis of environmental pollutants is a very important part of the decision-making process involved in the management of potentially contaminated site problems. All environmental samples that are intended for use in the characterization of potentially contaminated sites must therefore be collected, handled, and analyzed properly, in accordance with applicable quality control standards and methods. The following represents a particularly important checklist of items that should be reviewed in the development of a sampling and analysis workplan (CCME, 1993; Keith, 1988, 1991):

- What observations at sampling sites are to be recorded?
- Has information concerning data quality objectives, analytical methods, limits of detection, etc., been included?
- Have instructions for modifying protocols in case of problems been specified?
- Has a list of all sampling equipment and materials been prepared?
- Are there instructions for cleaning equipment before and after sampling?
- Have instructions for each type of sample collection been prepared?
- Have instructions for completing sample labels been included?
- Have instructions for preserving each type of sample (such as preservatives to use and also maximum holding times of samples) been included?
- Have instructions for packaging, transporting, and storing samples been included?
- Have instructions for chain-of-custody procedures been included?
- Have health and safety plans been developed?
- Is there a waste management plan to deal with investigation-generated wastes?

C.1.1 Sampling and Analysis Plan Checklist and Requirements

The following represents a convenient checklist of the issues that should be verified when planning a sampling activity for contaminated sites (CCME, 1993):

- What are the DQOs, and what corrective measures are planned if DQOs are not met (e.g. re-sampling or revision of DQOs)?
- Do program objectives need exploratory, monitoring, or both sampling types?
- Have arrangements been made for site entry or access?
- Is specialized sampling equipment needed and/or available?
- Are field crew who are experienced in the required types of sampling available?
- Have all analytes and analytical methods been listed?
- Have required good laboratory practice and/or method QA/QC protocols been listed.
- What type of sampling approach will be used (i.e. random, systematic, judgmental, or combinations thereof)?
- What type of data analysis methods will be used (e.g. geostatistical, control charts, hypothesis testing, etc.)?
- Is the sampling approach compatible with data analysis methods?
- How many samples are needed?
- What types of quality control (QC) samples are needed, and how many of each type of QC samples are needed (e.g. trip blanks, field blanks, equipment blanks, etc.)?

Furthermore, at a minimum, the following set of information should be provided for documenting the environmental sampling activities (CCME, 1993):

- Sampling date
- Sampling time
- Sample identification number
- Sampler's name
- Sampling site
- Sampling conditions or sample type
- Sampling equipment
- Preservation used
- Time of preservation
- Relevant sample site observations (auxiliary data).

Also, at a minimum, the following set of information should be provided for documenting the laboratory work performed to support site characterization activities (CCME, 1993; USEPA, 1989):

- Method of analysis
- Date of analysis
- Laboratory and facility carrying out analysis
- Analyst's name
- Calibration charts and other measurement charts (e.g. spectral)
- Method detection limits
- Confidence limits
- Records of calculations
- Actual analytical results.

C.2   Health and Safety Requirements for the Investigation of Contaminated Sites

The purpose of a health and safety plan (HSP) is to identify, evaluate, and control health and safety hazards, and to provide for emergency response during site characterization activities at a potentially contaminated site. A typical HSP will consist of the following general outline and elements:

- Introduction (consisting of site information and the identification of responsible personnel, such as the health and safety officer and an on-site safety manager)
- General safety requirements (including a discussion of safety requirements to be met, and the type of emergency equipment required)
- Employee protection program
- Decontamination procedures
- Health and safety training
- Emergency response (including emergency telephone numbers, personal injury actions, acute exposure to toxic materials responses, and directions to nearest hospital or medical facility).

In general, every project should start with a health and safety review (at which all site personnel sign a review form); a tailgate safety meeting (to be attended by all site activities personnel); and a safety compliance agreement (that should be signed by all persons entering the site, i.e both site personnel and site visitors). Contractors, subcontractors and other investigative teams are required to implement the HSP during site characterization and site restoration activities.

C.2.1   The Elements of a Site-specific Health and Safety Plan

The following represents the relevant elements of a HSP that will satisfy the general requirements of a safe work activity (CDHS, 1990; USEPA, 1987):

- Description of known hazards and risks associated with site activities (i.e. a health and safety risk analysis for existing site conditions, and for each site task and operation).
- Listing of key personnel and alternatives responsible for site safety, emergency response operations, and public protection.
- Description of the levels of protection to be worn by investigative personnel and visitors to the site.
- Delineation of work and rest areas.
- Establishment of procedures to control site access.
- Description of decontamination procedures for personnel and equipment.
- Establishment of site emergency procedures, including emergency medical care for injuries and toxicological problems.
- Development of medical monitoring program for personnel (i.e. medical surveillance requirements).
- Establishment of procedures for protecting workers from weather-related problems.
- Specification of any routine and special training required for personnel responding to environmental or health and safety emergencies.
- Definition of entry procedures for confined spaces.
- Description of requirements for environmental surveillance program.
- Description of the frequency and types of air monitoring, personnel monitoring, and environmental sampling techniques and instrumentation to be used.

It is important that these elements are fully evaluated, to ensure compliance with local health and safety regulations and/or to avert potential health and safety problems during site activities.

C.2.2   The Health and Safety Personnel

The responsible personnel assigned to see to the implementation of the site-specific HSP usually will include a Health and Safety Officer (HSO), and an Onsite Health and Safety Manager (OHSM); the responsibilities of the HSO and the OHSM may be assumed by the same individual.

The HSO has the primary responsibility for ensuring that the policies and procedures of the HSP are implemented by the OHSM. The HSO ensures that all personnel designated to work at the site:

- have been declared fit for the specific tasks by a physician or other qualified health professional,
- are able to wear air purifying respirators (should it become necessary), and
- would have received adequate hazardous waste site operations and emergency response training.

The HSO is also responsible for providing the appropriate safety monitoring equipment, and other resources necessary to implement the HSP. Significant deviations or changes to the original HSP must be approved by the HSO; occasionally, the OHSM will be given the authority to resolve outstanding health and safety issues that come up during site operations.

The OHSM supervises all site activities, and is responsible for implementing the HSP. The OHSM is responsible for providing copies of the HSP to the site crew (including subcontractors) and for advising the site crew on all health and safety matters. Specific tasks generally performed by the OHSM include the following:

- Inspection of site prior to start of work, especially for areas of concern that may have been omitted in the HSP or other areas that require special attention.
- Conducting daily safety briefings and site-specific training for on-site personnel prior to commencing work.
- Modifying or developing health and safety procedures, after consultation with the HSO, when site or working conditions change.
- Maintaining adequate safety supplies and equipment on site.
- Maintaining and supervising site control, decontamination, and contamination-reduction procedures.
- Investigating all accidents and incidents that occur during site activities.
- Disciplining or dismissing personnel whose conduct does not meet the requirements of the HSP, or whose conduct may jeopardize the health and safety of the site crew.
- Immediate notification of the emergency response authorities (e.g. the Fire and Police Departments), and the implementation of evacuation procedures during an emergency situation.
- Coordination of emergency response equipment, and also coordination of the transport of affected personnel to the nearest emergency medical care (including subsequent personnel decontamination), in the event of an emergency.

The OHSM has stop-work authority if a dangerous or potentially dangerous and unsafe situation exists at the site. Consequently, the field crew must notify the OHSM of any changes in site conditions (that may have an impact on the safety of personnel, the environment, or property) during the course of fieldwork activities.

C.2.3 Minimum Safety Requirements for Activities at Contaminated Sites

The following represents the minimum safety requirements to adopt during the investigation of potentially contaminated site problems:

- At least one copy of the HSP should be available at the site at all times.
- In general, fieldwork will preferably be conducted during daylight hours only. The OHSM will have to grant special permission for any field activities beyond daylight hours.
- Ideally, there should be at least two persons in the field/site at all times of site activities.
- Eating, drinking, and smoking should be restricted to a designated area, and all personnel should be required to wash their hands and face before eating, drinking, or smoking.

- Shaking off and blowing dust or other materials from potentially contaminated clothing or equipment should be prohibited.
- The OHSM should take steps to protect personnel engaged in site activities from potential contacts with splash water generated from decontamination of samplers, augers, etc.

In addition, the OHSM should take steps to protect personnel engaged in site activities from such physical hazards as falling objects (such as tools or equipment); tripping over hoses, pipes, tools or equipment; slipping on wet or uneven surfaces; insufficient or faulty protective equipment; insufficient or faulty tools and equipment; overhead or below-ground electrical hazards; heat stress and strain; insect and reptile bites; inhalation of dust, etc. The OHSM should also monitor and check the work habits of the site crew to ensure that they are safety-conscious. Decontamination and emergency procedures are also an important part of the overall employee protection program.

C.2.4   Specific Safety and Emergency Needs

A variety of safety and emergency equipment and supplies is generally required on-site at all times during the investigation of potentially contaminated sites. These may include the following suggested items:

- First aid kits containing bandaging materials (e.g. band aids, adhesive tape, gauze pads and rolls, butterfly bandages, and splints), anti-bacterial ointments, oxygen, pain killers (e.g. aspirin, acetaminophen), etc.
- Plastic sheetings
- Shovels and tools
- Warning tapes
- Personal protection equipment, including respirators, personnel protective suits, hard hats, goggles, boots, gloves, and ear-plugs (or other hearing protection apparatus required to minimize worker exposure to excessive and damaging noise from field activities such as during drilling operations)
- Potable water with electrolytic solution.

In general, the use of appropriate types of protection equipment will ensure the safety of the field crew. In all cases of contaminated site investigation programs, any additional precautions necessary to prevent exposure during drilling, sampling, and decontamination procedures should be implemented as site and weather conditions change. At all times, sufficient drinking water and safety equipment should be made available as necessary.

C.2.5   Emergency Response Issues

Several provisions for an emergency response plan should be carried out whenever there is a fire, explosion, or release of hazardous material which could threaten human health or the environment. The decision as to whether or not a fire, explosion, or hazardous material(s) release poses a real or potential hazard to human health or the environment is to be made by the OHSM.

To be able effectively to manage an emergency situation, emergency phone numbers should be compiled and included in the HSP. Also, the directions to the nearest hospital or medical facility, including a map clearly showing the shortest route from the site to the hospital or medical facility, should be kept with the HSP at the site.

C.3 Requirements for Investigation-generated Wastes

An IGW management plan, describing the storage, treatment, transportation, and disposal of any materials (both hazardous and non-hazardous) generated during a site characterization or site restoration activity, should be included in the project workplan. The most important elements of the IGW management approach are summarized as follows (USEPA, 1991):

- Characterize IGW through the use of existing information (e.g. manifests, Material Safety Data Sheets [MSDS], previous test results, knowledge of the waste generation process, and other relevant records) and best professional judgment.
- Leave a site in no worse condition than existed prior to the investigation or site activity.
- Remove from the site, those wastes that pose an immediate threat to human health or the environment.
- Delineate an 'area of contamination' unit for leaving on-site, wastes that do not require off-site disposal or extended above-ground containerization.
- Comply with all regulatory and legal requirements, to the extent practicable.
- Carefully plan and coordinate the IGW management program (e.g. containerize and dispose of hazardous groundwater, decontamination fluids, and personnel protection equipment at permitted facilities; *but* leave non-hazardous soil cuttings, groundwater, and decontamination fluids – preferably without containerization and testing – at the site of origination).
- Minimize the quantity of wastes generated during site activities.

To the extent practicable, the handling, storage, treatment, or disposal of any IGWs produced during site characterization and restoration activities must satisfy all regulatory requirements and stipulations that are applicable or relevant and appropriate to the site location. The procedures must also satisfy any limit requirements on the amount and concentration of the hazardous substances, pollutants, or contaminants involved.

C.4 Specification of QA/QC Samples

A detailed quality assurance/quality control (QA/QC) plan, describing specific requirements for QA and QC of both laboratory analysis and field sampling/analysis, should be part of the site characterization project workplan. The plan requirements will typically relate to, but not be limited to the following: the use of blanks, spikes, and duplicates; sampling procedures; cleaning of sampling equipment; storage; transportation; DQOs; chain-of-custody; and methods of analysis. In fact, some aspects of the field program should be subjected to a quality assessment survey, accomplished by submitting sample blanks (alongside the environmental samples) for analysis on a regular basis.

The various blanks and checks that are recommended as part of the quality assurance plan include the following (CME, 1994):

- *Trip blank*, required to identify contamination of bottles and samples during travel and storage. To prepare the trip blank, the laboratory fills containers with contaminant-free water and delivers them to the sampling crew; the field sampling crew subsequently ship and store these containers with the actual samples obtained from the site characterization activities. It is recommended to include one trip blank per shipment, especially where volatile contaminants are involved.
- *Field blank*, required to identify contamination of samples during collection. This is prepared in the same manner as the trip blank (i.e. the laboratory fills containers with contaminant-free water and delivers them to the sampling crew); subsequently, however, the field sampling crew expose this water to site air (just like the actual samples obtained from the site characterization activities). It is recommended to include one field blank per site or sampling event/day.
- *Equipment blanks*, required to identify contamination from well and sampling equipment. To obtain an equipment blank, casing materials and sampling devices are flushed with contaminant-free water, which is then analyzed. Typically, equipment blanks become important only if a problem is suspected, such as using a bailer to sample from multiple wells.
- *Blind replicates*, required to identify laboratory variability. To prepare the blind replicate, a field sample is split into three containers and labeled as different samples before shipment to the laboratory for analyses. It is recommended to include one blind replicate per day, or an average of one per 10 to 25 samples where large numbers of samples are involved.
- *Spiked samples*, required to identify errors due to sample storage and analysis. To obtain the spiked sample, known concentration(s) are added to the sample bottle and then analyzed. It is recommended to include one spiked sample per site, or an average of one per 25 samples where large numbers of samples are involved.

The development and implementation of a good QA/QC program during a sampling and analysis program is indeed critical to obtaining reliable analytical results for the site characterization.

C.5 References

CCME (Canadian Council of Ministers of the Environment) (1993). *Guidance Manual on Sampling, Analysis, and Data Management for Contaminated Sites.* Vol. I – *Main Report* (Report CCME EPC-NCS62E), and Vol. II – *Analytical Method Summaries* (Report CCME EPC-NCS66E). The National Contaminated Sites Remediation Program, Winnipeg, Manitoba, December 1993.

CCME (Candian Council of Ministers of the Environment) (1994). *Subsurface Assessment Handbook for Contaminated Sites.* Canadian Council of Ministers of the Environment (CCME), The National Contaminated Sites Remediation Program (NCSRP), Report No. CCME-PC-NCSRP-48E (March 1994), Ottawa, Ontario, Canada.

CDHS (California Department of Health Services) (1990). *Scientific and Technical Standards for Hazardous Waste Sites.* Prepared by the California Department of Health Services, Toxic Substances Control Program, Technical Services Branch, Sacramento, California.

Keith, L.H. (ed.) (1988). *Principles of Environmental Sampling*. American Chemical Society (ACS), Washington, DC.
Keith, L.H. (1991). *Environmental Sampling and Analysis – A Practical Guide*. Lewis Publishers, Boca Raton, Florida.
USEPA (US Environmental Protection Agency) (1987). RCRA Facility Investigation (RFI) Guidance, EPA/530/SW-87/001, Washington, DC.
USEPA (US Environmental Protection Agency) (1989). *User's Guide to the Contract Laboratory Program*. Office of Emergency and Remedial Response, OSWER Dir. 9240.0-1.
USEPA (US Environmental Protection Agency) (1991). *Management of Investigation-derived Wastes during Site Inspections*. Office of Emergency and Remedial Response. US Environmental Protection Agency, Washington, DC.

Appendix D

Major routes of potential receptor exposures to environmental contaminants

The analysis of potential receptor exposures to constituents present at a contaminated site often involves several complex issues related to the variety of contaminant migration and exposure pathways. This appendix discusses the receptor exposures for the different primary routes of contact, defined by the inhalation, ingestion, and dermal exposure relationships shown below and elaborated in greater detail elsewhere in the literature (e.g. CAPCOA, 1990; CDHS, 1986; DTSC, 1994; USEPA, 1988, 1989a, 1989b, 1991).

D.1 Inhalation Exposures

Two major types of inhalation exposure pathways are generally considered in the investigation of potentially contaminated site problems. The primary pathway is inhalation of airborne particulates from fugitive dust, in which all individuals within approximately 80 km ($\approx$ 50 miles) radius of a site are potentially impacted. A secondary exposure pathway relates to the inhalation of volatile compounds (i.e. airborne, vapor-phase chemicals).

Potential inhalation intakes are estimated based on the length of exposure, the inhalation rate of the exposed individual, the concentration of contaminant in the inhaled air, and the amount retained in the lungs. Potential receptor inhalation exposures to particulates from wind-borne fugitive dust releases and to volatile compounds from airborne vapor-phase emissions are shown in Boxes D.1 and D.2, respectively.

D.2 Ingestion Exposures

The major types of ingestion exposure pathways that could affect contaminated site management decisions consist of the oral intake of contaminated soils, food products, and waters.

Exposure through ingestion is a function of the concentration of the contaminant in the material ingested (e.g. soil, water, or food), the gastrointestinal absorption of the

---

Box D.1   Equation for estimating the inhalation exposure to contaminants in fugitive dust

$$INH_a = \frac{(C_a \times IR \times RR \times ABS_s \times ET \times EF \times ED)}{(BW \times AT)}$$

where:
- INH_a = Inhalation intake (mg/kg-day)
- C_a = Chemical concentration of airborne particulates (mg/m^3)
- IR = Inhalation rate (m^3/hr)
- RR = Retention rate of inhaled air (%)
- ABS_s = Percent of chemical absorbed into the bloodstream (%)
- ET = Exposure time (hours/day)
- EF = Exposure frequency (days/year)
- ED = Exposure duration (years)
- BW = Body weight (kg)
- AT = Averaging time (period over which exposure is averaged – days)
  = $ED \times 365$ days/years, for noncarcinogenic effects
  = 70 years $\times 365$ days/year, for carcinogenic effects

---

Box D.2   Equation for estimating the inhalation exposure to vapor-phase contaminants

$$INH_{qv} = \frac{(C_{av} \times IR \times RR \times ABS_s \times ET \times EF \times ED)}{(BW \times AT)}$$

where:
- INH_{qv} = Inhalation intake (mg/kg-day)
- C_{av} = Chemical concentration in air (mg/m^3)
- IR = Inhalation rate (m^3/hour)
- RR = Retention rate of inhaled air (%)
- ABS_s = Percent of chemical absorbed into the bloodstream (%)
- ET = Exposure time (hours/day)
- EF = Exposure frequency (days/year)
- ED = Exposure duration (years)
- BW = Body weight (kg)
- AT = Averaging time (period over which exposure is averaged – days)
  = $ED \times 365$ days/year, for noncarcinogenic effects
  = 70 years $\times 365$ days/year, for carcinogenic effects

---

pollutant in solid or fluid matrix, and the amount ingested. Potential receptor ingestion exposures through the oral intake of contaminated waters, through the consumption of contaminated food products, and through the incidental ingestion of contaminated soils/sediments are shown in Boxes D.3, D.4 and D.5, respectively.

D.3   Dermal Exposure

The major types of dermal exposure pathways that could affect contaminated site management decisions consist of dermal contacts with contaminants adsorbed on to

---

Box D.3   Equation for estimating the ingestion exposure to contaminated waters

$$ING_{dw} = \frac{(C_w \times WIR \times FI \times ABS_s \times EF \times ED)}{(BW \times AT)}$$

where:

ING_{dw}	=	Ingestion intake, adjusted for absorption (mg/kg-day)
C_w	=	Chemical concentration in water (mg/L)
WIR	=	Average water ingestion rate (liters/day)
FI	=	Fraction ingested from contaminated source (unitless)
ABS_s	=	Bioavailability/gastrointestinal (GI) absorption factor (%)
EF	=	Exposure frequency (days/years)
ED	=	Exposure duration (years)
BW	=	Body weight (kg)
AT	=	Averaging time (period over which exposure is averaged – days)
	=	$ED \times 365$ days/year, for noncarcinogenic effects
	=	70 years $\times$ 365 days/year, for carcinogenic effects

Box D.4   Equation for estimating the ingestion exposure to contaminated food products

$$ING_f = \frac{(C_f \times FIR \times CF \times FI \times ABS_s \times EF \times ED)}{(BW \times AT)}$$

where:

ING_f	=	Ingestion intake, adjusted for absorption (mg/kg-day)
C_f	=	Chemical concentration in food (mg/kg or mg/L)
FIR	=	Average food ingestion rate (mg or liters/meal)
CF	=	Conversion factor (10^{-6} kg/mg for solids and 1.00 for fluids)
FI	=	Fraction ingested from contaminated source (unitless)
ABS_s	=	Bioavailability/gastrointestinal (GI) absorption factor (%)
EF	=	Exposure frequency (meals/year)
ED	=	Exposure duration (years)
BW	=	Body weight (kg)
AT	=	Averaging time (period over which exposure is averaged – days)

Box D.5   Equation for estimating the incidental ingestion of contaminated soils/sediments

$$ING_s = \frac{C_s \times SIR \times CF \times FI \times ABS_s \times EF \times ED}{BW \times AT}$$

where:

ING_s	=	Ingestion intake, adjusted for absorption (mg/kg-day)
C_s	=	Chemical concentration in soil (mg/kg)
SIR	=	Average soil ingestion rate (mg soil/day)
CF	=	Conversion factor (10^{-6} kg/mg)
FI	=	Fraction ingested from contaminated source (unitless)
ABS_s	=	Bioavailability/gastrointestinal (GI) absorption factor (%)
EF	=	Exposure frequency (days/years)
ED	=	Exposure duration (years)
BW	=	Body weight (kg)
AT	=	Averaging time (period over which expoure is averaged – days)

Box D.6   Equation for estimating dermal exposures through contacts with contaminated soils/sediments

$$DEX = \frac{(C_s \times CF \times SA \times AF \times ABS_s \times SM \times EF \times ED)}{(BW \times AT)}$$

where:
DEX	= Absorbed dose (mg/kg-day)
C_s	= Chemical concentration in soil (mg/kg)
CF	= Conversion factor (10^{-6} kg/mg)
SA	= Skin surface area available for contact, i.e. surface area of exposed skin (cm^2/event)
AF	= Soil to skin adherence factor, i.e. soil loading on skin (mg/cm^2)
ABS_s	= Skin absorption factor for chemicals in soil (%)
SM	= Factor for soil matrix effects (%)
EF	= Exposure frequency (events/year)
ED	= Exposure duration (years)
BW	= Body weight (kg)
AT	= Averaging time (period over which exposure is averaged – days)

Box D.7   Equation for estimating dermal exposures through contacts with contaminated waters

$$DEX_w = \frac{(C_w \times CF \times SA \times PC \times ABS_s \times ET \times EF \times ED)}{(BW \times AT)}$$

where:
DEX_w	= Absorbed dose from dermal contact with chemicals in water (mg/kg-day)
C_w	= Chemical concentration in water (mg/L)
CF	= Volumetric conversion factor for water (1 liter/1000 cm^3)
SA	= Skin surface area available for contact, i.e. surface area of exposed skin (cm^2)
PC	= Chemical-specific dermal permeability constant (cm/hour)
ABS_s	= Skin absorption factor for chemicals in water (%)
ET	= Exposure time (hours/day)
EF	= Exposure frequency (days/year)
ED	= Exposure duration (years)
BW	= Body weight (kg)
AT	= Averaging time (period over which exposure is averaged – days)

soils and dermal absorption from contaminated waters. Dermal intake is determined by the chemical concentration in the medium of concern, the body surface area in contact with the medium, the duration of the contact, the flux of the medium across the skin surface, and the adsorbed fraction. Potential receptor dermal exposures through dermal contacts with contaminated soils/sediments and from the dermal absorption of chemicals present in contaminated waters are shown in Boxes D.6 and D.7, respectively.

D.4 References

CAPCOA (California Air Pollution Control Officers Association) (1990). *Air Toxics 'Hot Spots' Program. Risk Assessment Guidelines*. California Air Pollution Control Officers Association, California.

CDHS (California Department of Health Services) (1986). *The California Site Mitigation Decision Tree Manual*. California Department of Health Services, Toxic Substances Control Division, Sacramento, California.

DTSC (Department of Toxic Substances Control) (1994). *Preliminary Endangerment Assessment Guidance Manual* (A guidance manual for evaluating hazardous substance release sites). California Environmental Protection Agency, DTSC, Sacramento, California.

USEPA (US Environmental Protection Agency) (1988). *Superfund Exposure Assessment Manual*. Report No. EPA/540/1-88/001, OSWER Directive 9285.5-1, USEPA, Office of Remedial Response, Washington, DC.

USEPA (US Environmental Protection Agency) (1989a). *Exposure Factors Handbook*, EPA/600/8-89/043.

USEPA (US Environmental Protection Agency) (1989b). *Risk Assessment Guidance for Superfund*. Vol. I – *Human Health Evaluation Manual* (Part A). EPA/540/1-89/002. Office of Emergency and Remedial Response, Washington, DC.

USEPA (US Environmental Protection Agency) (1991). *Risk Assessment Guidance for Superfund*. Vol. I: *Human Health Evaluation Manual*. Supplemental Guidance. *Standard Default Exposure Factors* (Interim Final). March 1991. Office of Emergency and Remedial Response, Washington, DC. OSWER Directive: 9285.6-03.

Appendix E

Questionnaire for the qualitative evaluation of corrective action assessment and response programs

---

Questionnaire Chart for Corrective Action Assessment and Response Determination

- Do contaminant sources exist?
- Are there visible sources of contamination?
- What are the sources of the contaminants?
- Is the site potentially contaminated with hazardous or toxic chemicals? If so, what are the toxic agents involved?
- Are there confining layers or porous layers in the soil horizon?
- Is there soil erosion, or recent cuts or fills on site?
- What is the nature of drainage and surface flow patterns at the site and immediate vicinity?
- What are the site characteristics, hydrological features, meteorological or climatic factors, land-use patterns, and agricultural practices affecting the transport and distribution of site contaminants?
- What is the distribution of the chemicals over the site and vicinity?
- Are there known 'hot spots' at the site?
- What is an appropriate 'background' or 'control' region to use for corrective action investigations?
- Is there any area that poses an immediate and life-threatening exposure?
- What are the important transport processes and migration pathways that contribute to intermedia transforms and the spread of contamination and/or exposure?
- Are there current or future potential receptors that could be adversely affected by the contaminant sources? In particular, are there sensitive ecosystems or residences located downgradient, downstream, or downwind from the site?
- Are there one or more pathways through which site contaminants might migrate from the source and reach potential receptors? What are the dominant routes of exposure at the site?
- Have populations already been impacted, and/or are populations potentially at risk?
- What are the potential risks posed to human health and the environment if no further response action is taken at the site?
- Does the risk level exceed benchmark levels specified by environmental compliance regulations? If so, what site-specific cleanup criteria will be appropriate for the site?

(continued overleaf)

(continued)

- At the indicated contaminant concentrations at the site, which areas are considered as posing risks to the environment or surrounding populations? Which areas must therefore be remediated in order to reduce risks to an 'acceptable' level?
- Are estimated risk levels low enough such that a 'no-action' alternative is still protective of public health and the environment?
- What contaminants and environmental media should become the target for site remediation?
- How much contaminated material should be remediated to achieve an acceptable site restoration goal?
- Which remedial alternatives can be applied at the contaminated site in order to achieve adequate cleanup?
- Will exposure pathways be interrupted or will receptors be protected following a removal or remedial actions?
- What institutional control measures and risk management strategies are required in the overall corrective action decision?

Appendix F

Selected units and measurements

Mass/Weight Units

g	gram(s)
ton (metric)	tonne $= 1 \times 10^6$ g
kg	kilogram(s) $= 10^3$ g
mg	milligram(s) $= 10^{-3}$ g
μg	microgram(s) $= 10^{-6}$ g
ng	nanogram(s) $= 10^{-9}$ g
pg	picogram(s) $= 10^{-12}$ g

Approximate Mass Conversions

1 g	=	0.035 oz
1 ton	=	2205 lb
1 kg	=	2.25 lb
1 mg	=	10^{-3} g
1 μg	=	10^{-6} g
1 ng	=	10^{-9} g
1 pg	=	10^{-12} g

Volumetric Units

cc or cm^3	cubic centimeter(s) $= 10^{-3}$ L
mL	milliliter(s) $= 10^{-3}$ L
L	liter(s) $= 10^3$ cm^3
m^3	cubic meter(s) $= 10^3$ L

Approximate Volume Conversions

1 cc	=	1 mL
1 mL	=	10^{-3} L
1 L	=	0.88 liquid quart [UK] (or, 0.95 US quart)
m^3	=	35 cubic feet

Environmental Concentration Units

ppm   parts per million
ppb   parts per billion
ppt   parts per trillion

Concentration Equivalents

1 ppm $\equiv$ mg/kg or mg/L
1 ppb $\equiv$ μg/kg or μg/L
1 ppt $\equiv$ ng/kg or ng/L

Concentrations in soils or other solid media
mg/kg   mg chemical per kg weight of sampled medium

Concentrations in water or other liquid media
mg/L   mg chemical per liter of total liquid volume

Concentrations in air media
mg/m^3 mg chemical per m^3 of total fluid volume

Units of Chemical Intake and Dose

mg/kg-day = milligrams of chemical exposure per unit body weight of exposed receptor per day.

Typical Expressions Commonly used in Environmental Management Programs

'Order of magnitude'
Reference to each 'order of magnitude' means the base parameter may vary by a factor of 10. It is often used in reference to the calculation of environmental quantities or risk probabilities.

Exponentials denoted by 10^κ
Superscripts refer to the number of times '10' is multiplied by itself. For example, $10^2 = 10 \times 10 = 100$; $10^3 = 10 \times 10 \times 10 = 1000$; $10^6 = 10 \times 10 \times 10 \times 10 \times 10 \times 10 = 1\,000\,000$.

Exponentials denoted by $10^{-\kappa}$
Negative supserscript is equivalent to the reciprocal of the positive term, i.e. $10^{-\kappa}$ equals $1/10^\kappa$. For example, $10^{-2} = 1/10^2 = 1/(10 \times 10) = 0.01$; $10^{-3} = 1/10^3 = 1/(10 \times 10 \times 10) = 0.001$; $10^{-6} = 1/10^6 = 1/(10 \times 10 \times 10 \times 10 \times 10 \times 10) = 0.000\,001$.

Exponentials denoted by $X.YZE + \kappa$
Number after the 'E' indicates the power to which 10 is raised, and then multiplied by the preceding term (i.e. the number of times $^-10^\kappa$ is multiplied by preceding term, or $X.YZ \times 10^\kappa$). For example, $1.00E - 01 = 1.00 \times 10^{-1} = 0.1$; $1.23E + 04 = 1.23 \times 10^{+4} = 12\,300$; $4.44E + 05 = 4.44 \times 10^5 = 444\,000$.

Index

Note: Figures in *italic* are to glossary references.

Index prepared by Liz Granger